ACID RAIN

A Bibliography of Research Annotated for Easy Access

by

G. Harry Stopp, Jr.

The Scarecrow Press, Inc.
Metuchen, N.J., and London
1985

Library of Congress Cataloging in Publication Data

Stopp, G. Harry.
Acid rain.

Includes index.
1. Acid rain--Bibliography. I. Title.
Z5862.2.A26S76 1985 016.3637'386 85-10858
[TD196.A25]
ISBN 0-8108-1822-1

Manufactured in the United States of America

TABLE OF CONTENTS . . . .

ACKNOWLEDGMENTS . . . .

This book would not have been possible without the use of the facilities of the Library of Congress and the libraries and librarians of the University of North Carolina at Chapel Hill, North Carolina State University, Virginia Polytechnic Institute and State University, the University of Virginia, the EPA Research Facility at Research Triangle Park, North Carolina and the U.S. Geological Survey headquarters in Reston, Virginia. A very special debt is owed to Gaylor Callahan and Kelly Sink of the Interlibrary Loan Department of the Jackson Library on the campus of the University of North Carolina at Greensboro.

INTRODUCTION . . . .

When fossil fuels are burned, sulfur, nitrogen, and carbon are oxidized and these oxides are released in aerosol form, as smoke, or in more technical terms, effluent. These compounds are carried upward by the convection associated with heating of air by the combustion of whatever fossil fuel is burning, and they become part of the atmosphere. Because many of the oxide particles are heavier than the surrounding air, they will fall back to earth and settle as deposits on leaves, grass, lakes, rocks, and buildings. This material, usually called soot, can form significant deposits if enough combustion occurs in a given region over a long period of time; the sooty streets and façades of industrial towns and cities in England and Ireland are often given as an example of what can happen when large amounts of coal are burned.

Smaller particles of carbon, sulfur, and nitrogen oxide may linger in the air longer than will the large particles. These can be incorporated into the water droplet coalescence process and return to earth as precipitation. When this occurs, the oxides mix chemically with the atmospheric moisture and form weak sulfuric, nitric, and carbonic acids which are then an integral part of the precipitation. The term most often applied to this acidified precipitation is "acid rain."

Industrialized cities and industrial regions in Europe and North America have suffered the consequences of combustion byproduct deposition since the beginning of the Industrial Revolution. Residential fireplaces and coal-burning stoves were the initial culprits, benign devices by themselves but malignant polluters when clustered in the extreme densities created by high levels of urbanization. As steam power was introduced to manufacturing and steel-making became a major industry, industrial sites themselves became major sources of air-borne oxides which were then deposited, in dry or wet form, on the landscapes of industrial regions. Such regions began to suffer with blighted vegetation, acidic lakes and streams, and high levels of corrosion on buildings and statues because acid was falling from the sky every day. Copper Hills, Tennessee, a major smelter site, was almost completely denuded by local pollution and resultant acid rain; Wheeling, West Virginia, in which both urbanization and industrialization were active local polluters, recorded a rainfall in 1971 which had a pH 2.2, a level of acidity similar to that of common battery acid [pH is used to express relative acidity or alkalinity].

During the late 1960s and early 1970s, in an attempt to alleviate the pollution problems created by large-scale combustion of fossil fuels, industrialized countries began to develop a series of air pollution policies and plans for action. The United States was particularly active in attacking the problem of air pollution and the programs designed to regulate industrial (perhaps the term "major source" would be better) effluent took three general tacks:

1. Control of the material being burned--this includes switching from coal or oil to gas, which burns much cleaner; using low-sulfur coal when coal is necessary; and moving from fossil fuel to nuclear-powered electric generating plants. This coincided nicely with an international shortage of natural gas and the OPEC power plays which reduced the availability of fuel oil.
2. Cleaning the effluent from major sources by physical or chemical processes--introduction of "scrubbers" to industrial smokestacks.
3. Raising smokestack heights so the pollution plume can escape the local or regional atmosphere and be transported into the upper atmosphere where it can be naturally cleansed or dispersed into harmless concentrations. This was called the "tall stacks policy" by the U.S. Environmental Protection Agency.

The first two suggested tactics work well when implemented. There has been a significant reduction of air pollution from major combustion sources when relatively clean fuels are consumed and the technological advances in scrubber design that have increased their efficiency are in place. Political and economic considerations have limited implementation of these two policies somewhat but, on a national scale, they have been successful.

On a local or regional scale the tall stacks policy has also been a success. Smokestack extension is a relatively inexpensive antipollution action and, in cities with traditionally high levels of industrial pollution such as Pittsburgh, Gary, and Cleveland, the installation of tall stacks has been credited with major improvements in the air quality. Local and regional acid rain problems have almost disappeared. The "ultimate" tall stack may be the 1,200 foot smokestack built at a smelter site in Sudbury, Ontario (the Eiffel Tower is almost 300 feet shorter).

Unfortunately, the atmospheric portion of the closed planetary ecosystem did not simply absorb the pollution the tall stacks sent into it. Rather than "cleansing" itself harmlessly or dispersing the chemicals into meaningless concentrations, the atmospheric circulation simply transported the sulfur, nitrogen, and carbon compounds downwind, sometimes hundreds of miles. The air pollution created by Gary thus became acid rain in upstate New York, or rural Ontario, or pristine Nova Scotia.

Acid rain became a noticeable problem in eastern North America during the 1970s. The average pH of precipitation in the eastern United States fell from a normal value of 5.7 (measured by NOAA for 30 years) to an acidic 4.2 by 1975. Single incident levels of 3.3 were becoming commonplace in the Adirondack Mountains and along the coast of Nova Scotia; neither region is industrialized but both are downwind in the general upper atmosphere circulation from such Industrial Heartland cities as Gary, Pittsburgh, Toledo, and Cleveland.

Canadian scientists began bringing the acid rain phenomenon to our attention because Canadian forests and lakes were feeling the impact of acid rain which deposited combustion effluent created in the United States, but U.S. policymakers did not listen. As ecologists in New York and New Hampshire began to discover acid-induced fish kills in pristine forest locations, local and then regional public and scientific concern grew. Precipitation pH is lowering in all states east of the Mississippi and in many regions in the Midwest and West.

Emission projections made in Global 2,000, the 1981 report by the President's Council for Environmental Quality [CEQ], indicate that the trend toward further acidification

Precipitation pH and Effects

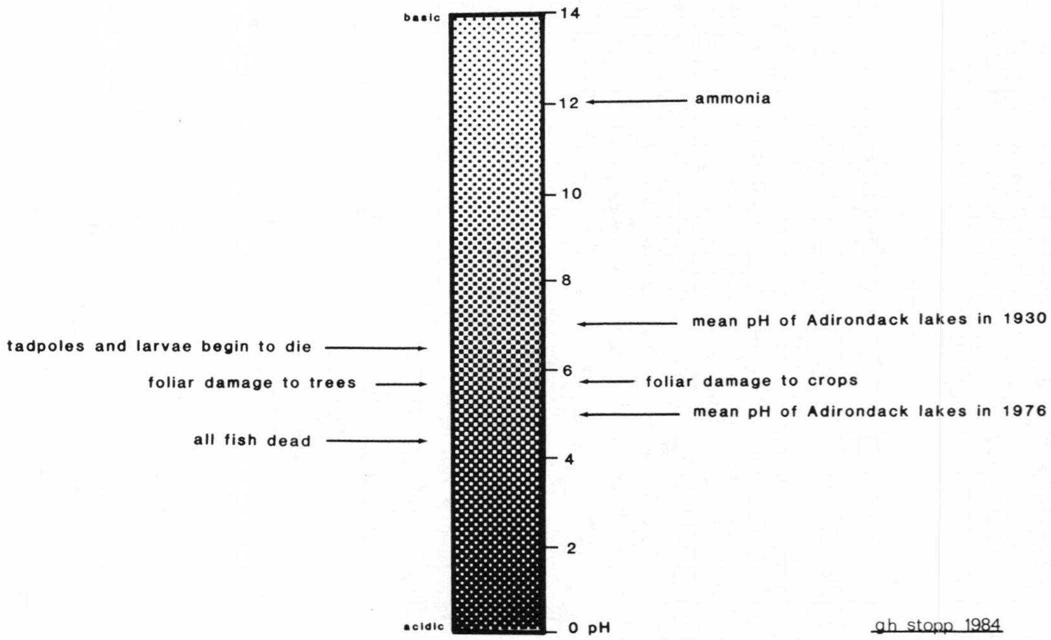

of the environment will continue. The figures provided by CEQ are:

EMISSION PROJECTIONS FOR THE UNITED STATES (millions of short tons)

	1985	1990
Sulfur dioxide	14.2	15.6
Oxides of nitrogen	17.1	18.2

The United States government has launched several national programs of research and data collection in an attempt to understand the extent and level of acid rain impact. It has also begun support of research into the effects of acid rain on various segments of the environment, from trout in a native stream to materials used in construction. The U.S. Department of Agriculture is initiating a national program to study the effects of acid rain on all agricultural activities. The U.S. Department of State along with the Department of Energy and the Environmental Protection Agency are engaged in discussions with the Canadian government to determine what if any international steps can be taken to control acid rain or to counter its effects. Only two days before I wrote this sentence, the Prime Minister of Canada and the President of the United States met to discuss the creation of a bilateral task force to focus on the problems created by acid rain.

As a scientist, I believe that the key to understanding is research and that research should be shared and discussed with my colleagues and with interested laypersons. Acid rain is a phenomenon that demands scientific attention and the immediacy of the problems created by the effects of acid rain require that scientists, policy makers, and the public obtain as much information about the causes and effects of acid rain as is possible in as short a time as is possible.

One of the problems with communication between scientists and non-scientists is that the two groups do not write to or talk to each other. Articles in scientific journals that can have a direct bearing on political or regulatory decisions may never become a factor in policy decisions because policy makers are seldom scientists and do not read scientific journals. Scientists, who do not generally read public administration or policy-oriented journals, may not be aware of ongoing policy development processes that will have a direct bearing on their research interest.

This bibliography is intended to offer scientists, policy-makers, and the general public a reference source that can help them locate useful information about acid rain, whatever their interest. The references listed herein include articles and books written about acid rain and, in a few cases, about subjects that directly relate to an understanding of acid rain and its effects. It does not, however, include newspaper articles, popular magazine articles or, with a few exceptions, unpublished theses or dissertations. Government documents have been screened out when they were determined to serve only bureaucratic purposes. No single aspect of the acid rain phenomenon is intended to dominate the references included; agronomy, atmospheric physics, biology, economics, geography, limnology and political science are all included in this compilation. This is intended to be a reference book for members of all disciplines, including interested laypersons.

Each of the entries is annotated. The annotations are to help the reader decide whether or not a particular entry is of use to his or her project. The reader will not find a synopsis of each book or article in the annotations but will be given information about the general thrust of an entry and about the skill or educational level needed to comprehend each entry.

A complete listing of sources is provided with the thought that an interested reader would want to scan a comprehensive list the same way one would browse through the stacks in a library. We often do not know what we are looking for until we see it and, if we only look in places we know about, we will never discover that special book or article that isn't categorized just the way we might have thought. A subject index is provided to help you reach a particular point of interest more quickly.

Some of the references are relatively obscure, particularly those that come from proceedings of conferences and workshops. Because the study of acid rain is a very recent undertaking, much seminal material is available only through such sources and I have tried to create bibliographic entries that make locating these sources as efficient as possible. In some cases, this has meant taking literary license with bibliographic style, an institution that frowns on license of any kind. I apologize to bibliographers but need not do so to potential users.

The references included here reflect what could be compiled through end-of-year, 1984. The national rush to provide information about acid rain should increase the rate at which articles and books on this subject are published and, by the time this book reaches the reader, there may be two or three times as many entries available. I certainly hope that situation occurs and that the problem of acid rain fades into history as an interesting incident in which humankind overcame a potentially catastrophic environmental event by logical, compassionate, public action. A second edition of this bibliography will happily chronicle those events.

ACRONYMS AND INITIALISMS . . . .

A.E.R.E.	Administration for Energy Resources and the Environment
AESC	Atmospheric Environmental Service of Canada
AIChE	American Institute of Chemical Engineering
APCA	American Physical Chemistry Association
APIOS	Acidic Precipitation in Ontario Study
ASAP	American Survey of Atmospheric Pollution
BLM	Bureau of Land Management
CANSAP	Canadian Network for Sampling Precipitation
CAPMON	Canadian Air and Precipitation Monitoring Network
CEC	Canadian Environmental Council
CEQ	Council on Environmental Quality
EPA	Environmental Protection Agency
EPRI	Electric Power Research Institute
ERIC	Educational Research Information Center
GAO	General Accounting Office
GLAD	Great Lakes Atmospheric Deposition Network
ILWAS	Integrated Lake-Watershed Acidification Study
INCO	International Nickel Company Ltd.
ITFAP	Interagency Task Force on Acid Precipitation
JAPCA	Journal of the American Physical Chemistry Association
LRTAP	Long Range Transport of Air Pollutants
MAP3S	Multi-state Atmospheric Power Production Pollution Study
NAAQS	National Ambiant Air Quality Standards
NADP	National Atmospheric Deposition Program
NAPAP	National Acid Precipitation Assessment Program
NASA	National Aeronautics and Space Administration
NCAR	National Center for Atmospheric Research
NCLAN	National Crop Loss Assessment Network
NCSU	North Carolina State University
NDAP	National Data Assessment Program
NERBC	National Energy Resource Board of Canada
NOAA	National Oceanic and Atmospheric Administration
NPS	National Precipitation Survey
NRCC	Natural Resource Communication Center
NREL	Natural Resource Ecology Laboratory
NTN	National Trends Network
OECD	Organization of Economic Cooperation and Development
SAES	State Agricultural Experiment Stations

SNSF	Swedish National Scientific Foundation
TAPPI	Technical Association of the Pulp and Paper Industry
TVA	Tennessee Valley Authority
UAPSP	Utility Acid Precipitation Study Program
USDA	U.S. Department of Agriculture
USGS	U.S. Geological Survey

WHAT HAVE THEY DONE TO THE RAIN . . . .

Malvina Reynolds

Just a little rain falling all around
The grass lifts its head to the heavenly sound,
Just a little rain, just a little rain
What have they done to the rain?

Just a little boy standing in the rain,
The gentle rain that falls for years.
And the grass is gone, the boy disappears,
And rain keeps falling like helpless tears,
And what have they done to the rain?

Just a little breeze out of the sky,
The leaves pat their hands as the breeze blows by,
Just a little breeze with some smoke in its eye,
What have they done to the rain?

Just a little boy standing in the rain,
The gentle rain that falls for years.
And the grass is gone, the boy disappears,
And rain keeps falling like helpless tears,
And what have they done to the rain?

ACID RAIN . . . .

1. Abelson, P.H. "Acid Rain," Science, Vol. 221, no. 4606, 1983, p. 215.

A very brief comment about the phenomenon in general and the controversy, national and international, surrounding it. Of use more as insight into the growing public awareness of acid rain than as scientific data or analysis.

2. Abrahamsen, G. and G.J. Dollard. "Effects of Acid Precipitation on Forest Vegetation and Soil," in Ecological Effects of Acid Precipitation. Palo Alto, CA: Electric Power Research Institute, 1979, pp. 3-8.

An overview of what was known at the time of the effects, direct and indirect, of acid deposition in the forest environment. A pro-industry bias across this publication dictates that more questions and gray areas are exposed than answers provided. A good source of general information and a good starting place for potential research topics.

3. Abrahamsen, G. "Acid Precipitation, Plant Nutrients and Forest Growth," in Proceedings of the International Conference on Ecological Impacts of Acid Precipitation. Sandefjord, Norway: SNSF Project, 1980, pp. 58-63.

An overview of the ways in which acid rain can or might affect forest ecology by interfering with the natural round of organic decay and mineral transportation through forest soils. This work draws from international research and, as such, provides a comprehensive introduction to this subject.

4. _______, et al. "Effects of Artificial Acid Rain and Liming on Soil Organisms and the Decomposition of Organic Matter," in T.C. Hutchinson and M. Havas (eds), Effects of Acid Precipitation on Terrestrial Ecosystems. New York: Plenum Press, 1980, pp. 341-362.

Report on experiments with decomposition of organic matter in the upper layers of the soil when exposed to acid rain. The basic information on humus formation under these conditions is important but the real contribution here is the report on how the effects of acid rain in the soil can be combatted by adding lime to the soil. A landmark article.

5. Ackerman, B.A. and W.T. Hassler. "Beyond the New Deal: Coal and the Clean Air Act," Yale Law Journal, Vol. 89, no. 8, 1980, pp. 1466-1571.

An analysis of the legal and policy implications of enforcing and expanding regulation of high sulfur coal use in the United States. Good general policy analysis reference.

6. _______. Clean Coal/Dirty Air: How the Clean Air Act Became a Multi-billion Dollar Bail-out for High-sulphur Coal-Producers and What Should Be Done About It. New Haven: Yale University Press, 1981.

A journalistic analysis of the effects of the Clean Air Act on the coal industry in the U.S. Topic is sensational but not particularly well documented or written.

7. Adams, R.M. and T.D. Crocker. "Dose-Response and the Value of Information: The Case of Acid Deposition," in T.D. Crocker (ed), Economics of Acid Deposition. Ann Arbor, MI: Ann Arbor Science Publishers, 1982, pp. 46-59.

Rather than providing information about acid rain, this article focuses on the problems involved with obtaining and using data gathered by the various acid rain monitoring programs then active. Particularly negative toward the work done by the utility-supported groups. An important article for statistical research.

8. _______. "Economically Relevant Response Estimation and the Value of Information: Acid Deposition," in T.D. Crocker (ed), Economic Perspectives on Acid Deposition Control. Boston: Butterworth Publishers, 1984, pp. 35-64.

Deals with the use of biologically derived parameters in economic analysis. Economic models for decision-making are tested with the kind of data available from acid rain research performed by biologists and other natural scientists.

9. Alexander, M. "Effects of Acidity on Microorganisms and Microbial Processes," in T.C. Hutchinson and M. Havas (eds), Effects of Acid Precipitation on Terrestrial Ecosystems. New York: Plenum Press, 1980, pp. 363-374.

Very basic research into the effects of pH on the life cycles of important microorganisms that affect the soil and plant growth cycles. This brings together specific research reported by earlier researchers and is a good synthesis for agronomists or ecologists.

10. Alexander, T. "Some Burning Questions About Combustion," in R.R. Campbell and J.L. Wade (eds), Society and Environment: The Coming Collision. Boston: Allyn and Bacon, 1972, pp. 135-144.

A rather general article on air pollution problems, including acid rain, which result from hydrocarbon combustion. In addition to acid rain, the greenhouse effect is also analyzed. Examines some alternatives to energy production through combustion. Policy oriented.

11. Allison, W.R. and H.H. Harvey. "Methods for Assessing the Benthos of Acidifying Lakes," in R. Singer (ed), Effects of Acidic Precipitation on Benthos. Springfield, IL: North American Benthological Society, 1981, pp. 1-13.

A thorough examination of currently acceptable methodology for use in limnological investigations. While some of the techniques are generic, the overall methodological model is one designed specifically for acidic situations. Very useful for ecologists and biologists.

12. Almer, B.W., et al. "Effects of Acidification on Swedish Lakes," Ambio, Vol. 3, 1974, pp. 30-36.

A benchmark report on the way in which acid rain has changed the chemistry and ecology of lakes in Sweden. The data cover effects throughout the lacustrine profile and include information on the biotic evolution that can occur in acidified lakes as well. This is a must for any limnological investigation.

13. _______. "Sulfur Pollution and the Aquatic Ecosystem," in J.O. Nriagu (ed), Sulfur in the Environment: Ecological Impacts. New York: John Wiley and Sons, 1978, pp. 273-311.

A very scientific look at the ways in which sulfur enters and affects aquatic ecology. Not restricted to introduction of sulfur through acidic precipitation but does treat the subject.

14. Alsop, W. and D. DuBay. "Acid Rain in North Carolina: Deposition Monitoring and Effects Research," in J. Hernandez, (ed), A Healthy Economy in a Healthy Environment. Chapel Hill, NC: University of North Carolina Environmental Studies Council, 1983, pp. 402-409.

An overview of preliminary attempts at acid deposition monitoring in a program concerned with botanical reactions to acid rain. It does provide useful information for the field-oriented researcher because it can serve as a negative model.

15. Altshuller, A.P. and G.A. McBean. The LRTAP Problem in North America: A Preliminary Overview, report by the United States-Canada Bilateral Research Consultation Group on the Long-Range Transport of Air Pollutants. Downsview, Ontario: Atmospheric Environment Service, 1980.

Basically an administrative report on the status of an international commission. The progression of decision-making and policy development outlined in this report can serve as a valuable source of information to political scientists and public administration students.

16. American Chemical Society. Acid Rain: An Information Profile. Washington, DC: American Chemical Society, 1982.

A very general primer in encyclopedic form. This can serve a beginning or less technical reader very well.

17. Amthor, J.S. "Does Acid Rain Directly Influence Plant Growth: Some Comments and Observations," Environmental Pollution, Vol. 36, no. 1, 1984, pp. 1-6.

Discusses reported research on plant reaction to acid rain with an analytic approach. The author does not recognize any evidence of direct effects on plants from acid rain but does for indirect effects. He also puts forth the idea that other factors, perhaps related to acid rain, may be at fault, especially in cases of forest decline.

18. Anderson, R.J. and T.D. Crocker. "The Economics of Air Pollution: A Literature Assessment," in P.B. Downing (ed), Air Pollution and the Social Sciences. New York: Praeger Publishers, 1971, pp. 133-165.

An evaluation of the trends in scientific literature about the economics of air pollution. Identifies major theoretical biases and pro and con regulatory adjustment writings. A good overview for economists and policy analysts.

19. Andren, A.W. and S.E. Lindberg. "Atmospheric Input and Origin of Selected Elements in Walker Branch Watershed, Oak Ridge, Tennessee," Water, Air, and Soil Pollution, Vol. 8, 1977, pp. 199-215.

One of the first reports of acidic precipitation in the Southeast region of the U.S. Limited to a very small sub-region but baseline data for any further study.

20. Andresen, A.M., et al. "Levels of Lead, Copper, and Zinc in the Forest Floor in the Northeastern United States," Journal of Environmental Quality, Vol. 9, no. 2, 1980, pp. 293-296.

A comprehensive look at the way in which some of the important micronutrients in a forest ecosystem react to environmental changes, especially to acid rain and the resultant soil and soil moisture pH changes. This can serve as a seminal work to soil scientists, forest ecologists, and agronomists.

21. Applegate, H.G. and C. R. Bath (eds). Air Pollution Along the United States-Mexico Border. El Paso, TX: Texas Western Press, 1974.

A comprehensive analysis of pollution-related problems along the southern U.S. border and a very thorough evaluation of the development of international policy with regard to pollution control along that border. Provides a good political history also.

22. Applegate, H.G. "Transfrontier Air Pollution Along the United States-Mexico Border," The Environmentalist, Vol. 4, 1984, pp. 219-227.

An update of material presented in the author's earlier book on this subject. This traces twelve years of international negotiation about pollution policy that has local, regional, and international impact. Very useful to compare with the U.S.-Canada pollution policy developments.

23. Archer, R.J., et al. Basin Planning Report EMB-4, precipitation data report by the Water Resources Commission. Albany, NY: New York State Conservation Department, 1968.

A comprehensive collection of data collected on a state-wide basis. There is very little analysis or evaluation of the data presented but the information can be very useful for comparative study with other similar regional reports.

24. Arey, D. and J. Crenshaw. "The Management of the Potential Air Pollution Hazard from Stationary Sources," in J.W. Frazier and B.J. Epstein (eds), Applied Geography Conferences, Vol. 2. Binghamton, NY: State University of New York, 1979, pp. 213-224.

A rather weak attempt to develop a mathematical model for predicting environmental hazards created by industrial combustion sites. Illinois examples are used in this study but the article spends much more time on explaining the mathematical machinations than in developing or offering a usable explanation of the phenomenon discussed. Of limited utility to atmospheric scientists.

25. Arnold, D.E., et al. "Studies on Infertile, Acidic Pennsylvania Streams and Their Benthic Communities," in R. Singer (ed), Effects of Acidic Precipitation on Benthos. Springfield, IL: North American Benthological Society, 1981, pp. 15-33.

A report on benthic data collected over time in several highly acidified small streams in the Northeastern U.S. Good source of data on population dynamics among benthic species in an acidic environment. The specific focus on deep water species helps make this report almost unique and very useful.

26. ________. Vulnerability of Selected Lakes and Streams in the Middle Atlantic States to Acidification, Technical Report DBS-80/40.17. Washington, DC: U.S. Fish and Wildlife Service, 1984.

A thorough examination of the geologic, hydrologic, and climatic characteristics of certain lakes and streams which have an effect, either positive or negative, on their ability to withstand acid rain. Very good predictive models developed to help apply the knowledge gained from this study to lakes or streams in other regions.

27. Asako, K. "Environmental Pollution in an Open Economy," Economic Review, Vol. 29, 1979, pp. 359-367.

Presents some of the basic assumptions made and the problems created when "rampant capitalism" faces a society with real concerns for environmental quality. Without actively choosing sides, Asako introduces the reader to real problems in environmental economics and provides mechanisms that might be appropriate to begin solving these problems.

28. ________. "Economic Growth and Environmental Pollution Under the Max-Min Principle," Journal of Environmental Economics Management, Vol. 13, 1980, pp. 157-183.

A more technical analysis of the series of environmental and economic trade-offs involved with industrial and economic expansion in a capitalistic system. He attempts to present quantitative continua for specific situations of confrontation between the two spheres. Recommended for economists only.

29. Ashenden, T.W. and M. Mansfield. "Extreme Pollution Sensitivity of Grasses When SO_2 and NO_2 are Present in the Atmosphere Together," Nature, Vol. 273, 1978, pp. 142-143.

A research note on experiments with more-than-additive and less-than-additive situations.

30. Ashenden, T.W. "The Effects of Long-term Exposure to SO_2 and NO_2 Pollution on the Growth of Dactylis glomerata L. and Poa pratensis L.," Environmental Pollution, Vol. 18, 1979, pp. 249-258.

A rather technical but thorough report of controlled experiments with two agricultural plants. While the tests were on the exposure to gases, the implications for acidic precipitation containing these two gases are very clear and direct. A good article for botanists.

31. ________ and I.A.D. Williams. "Growth Reductions in Lolinum multiflorum and Phelum pratense as a Result of SO_2 and NO_2 Pollution," Environmental Pollution, Vol. 21, 1980, pp. 131-139.

A report of a similar set of experiments with two different plants as reported above. The results are mixed but can have implications for overall plant health in the presence of these two pollutants.

32. Ashmore, M.R., et al. "Visible Injury to Crop Species by Ozone in the United Kingdom," Environmental Pollution, Vol. 21, 1980, p. 209.

A short research report of agricultural effects by an acid-rain-related phenomenon. Particularly useful because ozone and acidic precipitation levels often coincide in industrial regions.

33. Asman, W.A.H. and P.J. Jonker. "Neutralization of Acid in Precipitation and Some Results of Sequential Rain Sampling," in H.W. Georgii and J. Pankruth (eds), Deposition of Atmospheric Pollutants. Boston: Reidel Publishing, 1982, pp. 115-123.

Primarily research methodology. Examines the high variability of acidity within specific rain events and how that can be accounted for by proper sampling techniques. Also examines the influence of continental and maritime air masses on acidic deposition.

34. Asman, W.A.H., et al. "Influence and Prevention of Bird-droppings in Precipitation Chemistry Experiments," Water, Air, and Soil Pollution, Vol. 17, 1982, pp. 415-420.

A must-read article outlining the problems associated with precipitation monitoring for acid rain experiments. The "bottom line" is build a bird ring when you need clean precipitation samples.

35. Ashworth, J.R. "Atmospheric Pollution and the Deposit Gauge," Weather, Vol. 3, 1941, pp. 237-240.

Basically a methodological note on the problems associated with using a standard rain gauging device in collecting samples for chemical analysis. Very old article but should be heeded.

36. Atkinson, S. "Effect of Global Optimization on Locally Optimal Pollution Control: Acid Rain," in T.D. Crocker (ed), Economic Perspectives on Acid Deposition Control. Boston: Butterworth Publishers, 1984, pp. 21-34.

The basic premise here is that "least cost strategies for SO_2 control that meet only local standards will most likely lead to increased long-range SO_4 deposition." This is basically a restatement of the well-known fact that the tall stacks policy established for local pollution control by EPA has created downwind acid rain problems. For those who need quantitative models of simple statements, both local diffusion and long-range transport models are provided by the author.

37. Aylor, D.E. "Deposition of Particles in a Plant Canopy," Journal of Applied Meteorology, Vol. 14, 1975, pp. 52-57.

A "pre-acid rain" article examining the general processes involved with deposition of aerosol particles on plant surfaces. Does present some of the possibilities for damage that pollutant deposition can create. Good basic scientific information on the deposition process.

38. Baath, E., et al. "Effects of the Artificial Acid Rain on Microbial Activity and Biomass," Bulletin of Environmental Contamination and Toxicology, Vol. 23, 1979, pp. 737-740.

Short outline of basic cycles and processes involved in the humus layer under acidification. Does provide a good set of data on overall plant productivity (biomass) under acid attack; useful for comparative study.

39. _______. "Soil Organisms and Litter Decomposition in a Scots Pine Forest--Effects of Experimental Acidification," in T.C. Hutchinson and M. Havas (eds), Effects of Acid Precipitation on Terrestrial Ecosystems. New York: Plenum Press. 1980, pp. 375-380.

A follow-up of the 1979 article by Baath [item 38] on the same subject. Very little new data.

40. Babish, H. and D.L. Davis. "Acid Rain: Causes and Consequences," Environment, Vol. 22, no. 4, 1980, pp. 6-13.

A general overview of the acid rain phenomenon. A good synthesis of what was commonly known in 1980 but in need of an update if utilized extensively now.

41. Bache, B.W. "The Sensitivity of Soils to Acidification," in T.C. Hutchinson and M. Havas (eds), Effects of Acid Precipitation on Terrestrial Ecosystems. New York: Plenum Press, 1980, pp. 569-572.

An overview of the way in which soil makeup can affect the way in which that soil reacts to acid rain. Good general analysis.

42. Baker, J.P. Effects on Fish of Metals Associated With Acidification. Bethesda, MD: American Fisheries Society, 1982.

A rather technical report on the association of heavy metal incidence with acidification. The relationship is further examined here by analysis of data on metal concentrations in fish populations in acidified and non-acidified situations. This report provides pretty substantial evidence of the direct correlation between acidification and increased metal pollution. Very good source of data as well as analytic models.

43. Ball. J.G. and W.R. Menzies. Flue Gas Desulfurization Cost Estimates for Acid Rain Studies: Technical Report, DCN 81-203-001-12-23. Washington, DC: Radian Corporation, 1981.

A very econometric report of cost estimates for installing various levels of pollution control technology in coal-burning industries. Very specialized information but critical to pollution control engineers.

44. Bangay, G.E. and C. Riordan (eds). Memorandum of Intent on Transboundary Air Pollution: Final Report. Washington, DC: U.S. Environmental Protection Agency, 1983.

A comprehensive (over 1,000 pages) collection of information concerning all aspects of air pollution in North America. Basically the technical backup for an international policy on acid rain. A real reference work.

45. Banwart, W.L., et al. "Acid Rain and Its Effects on Corn and Soybean Yields," Proceedings of the Illinois Fertilizer and Chemical Dealers Conference. Urbana, IL: University of Illinois, 1983, pp. 19-21.

A good, concise report of field and laboratory research on agricultural effects of acid rain. While the data act to substantiate earlier findings on soybeans, the information about corn is unique. Very useful for any study of agricultural effects.

46. Barnes, C.R. "Changes in the Chemistry of Bulk Precipitation in New York State, 1965-1978," Northeastern Environmental Science, Vol. 1, no. 3-4, 1982, pp. 187-197.

A good regional report on precipitation monitoring in one of the more acid-sensitive regions in the United States. Good source of data that can be used to supplement or complement that provided by national networks of data gathering. Interpretation and analysis is written for a general informed audience.

47. Baron, J. "Comparative Water Chemistry of Four Lakes in Rocky Mountain National Park," Water Resources Bulletin, Vol. 19, 1983, pp. 897-902.

Because acid rain is primarily an eastern U.S. problem, there are fewer data available for western environments. This research report, which is rather technical but very thorough, is particularly important because it gives some data on a western site and because that site is in a pristine environment. The effects are not very severe but the implications for further lacustrine influence are clear.

48. Barrett, E. and G. Brodin. "The Acidity of Scandinavian Precipitation," Tellus, Vol. 7, 1955, pp. 251-257.

One of the earliest comprehensive reports of acidification of precipitation in Europe. It can be used as a benchmark for comparative study and analysis.

49. Barrie, L.A. and J.L. Walmsler. "A Study of Sulphur Dioxide Deposition Velocities to Snow in Northern Canada," Atmospheric Environment, Vol. 12, 1978, pp. 2321-2332.

A scientific report on the special physical and chemical reactions involved with acidification of snow. Rather technical for the non-scientific reader but worthwhile if interested in a comprehensive understanding of atmospheric acidification.

50. Barrie, L.A. "The Prediction of Rain Acidity and SO_2 Scavenging in Eastern North America," Atmospheric Environment, Vol. 15, 1981, pp. 31-41.

An excellent development of atmospheric chemistry models of action and interaction with the goal of predicting downwind precipitation acidity. Rather technical but well written.

51. Barrie, L.A. and A. Sirois. An Analysis of Precipitation Chemistry Measurements Made by CANSAP: 1977-1980, Report AQRB-82-003-T, Atmospheric Environment Service. Ottawa: Environment Canada, 1982.

A most informative statistical analysis of the extensive data collected in a national monitoring program established by the Canadian government. The information on geographic and temporal trends in acidification are necessary for a really comprehensive understanding of the acid rain problem in North America.

52. Barrie, L.A. "Environment Canada's Long-range Transport of Atmospheric Pollutants Programme: Atmospheric Studies," in F.M. D'Itri (ed), Acid Precipitation: Effects on Ecological Systems. Ann Arbor, MI: Ann Arbor Science Publishers, 1982, pp. 141-161.

A fine evaluation of the predictive models Canada has developed to cope with international acid rain problems. Technical, statistical, but useful for scientific analysis.

53. Barrow, N.J. "Modelling the Effects of pH on Phosphate Sorption by Soils," Journal of Soil Science, Vol. 35, 1984, pp. 283-297.

A rather technical report on the observation of changes in uptake and retention of certain salts by soils under different pH regimes. Good background information for discussion of ionic dynamics and of soil structure.

54. Bath. C.R. "Alternative Cooperative Arrangements for Managing Transboundary Air Resources Along the Border," Natural Resources Journal, Vol. 18, 1977, pp. 197-198.

A description and analysis of proposed air pollution policies for control and regulation of pollution along the U.S.-Mexico border. Looks at alternatives below and above the national level and may provide insight into a possible solution for U.S.-Canadian policy.

55. Beamish, R.J. and H.H. Harvey. "Acidification of the La Cloche Mountain Lakes and Resulting Fish Mortalities," Journal of the Fisheries Research Board of Canada, Vol. 29, no. 8, 1972, pp. 1131-1143.

Provides a very clear insight into the effects of acid rain on fish resources in fresh water. The data given here have become a universal point of comparison for any study of lacustrine effects.

56. Beamish, R.J. "Loss of Fish Populations from Unexploited Remote Lakes in Ontario, Canada as a Consequence of Atmospheric Fallout of Acid," Water Resources, Vol. 35, no. 5, 1974, pp. 85-95.

A further exploration of the effects of acid rain on lacustrine resources. The results of the study of pristine lakes in Canada are frightening and have become the basis for much of the policy proposed by both Canadians and Americans for control of acid rain.

57. _______. "Growth and Survival of White Suckers (Catostomus commersoni) in an Acidified Lake," Journal of Fisher Resources Board of Canada, Vol. 31, 1974, pp. 49-54.

A good technical research report on a study of one species in Canadian water. Because it is a bottom-feeder, this fish is often used as an indicator for pollution and the data are very useful for comparative study.

58. _______, et al. "Long-term Acidification of a Lake and Resulting Effects on Fishes," Ambio, Vol. 4, 1975, pp. 98-102.

Technical research report on longitudinal study in Canada. A real benchmark study, especially valuable because of its early publication.

59. _______. "Acidification of Lakes in Canada by Acid Precipitation and the Resulting Effects on Fishes," Water, Air, and Soil Pollution, Vol. 6, 1976, pp. 501-514.

Basically a restatement of information presented in the two articles [items 57, 58]. The "new" information presented here is more policy- or general audience-oriented and, as such, of interest to a wider group of readers.

60. _______, et al. A Fish and Chemical Study of 109 Lakes in the Experimental Lakes Area (ELA), Northwestern Ontario, With Appended Reports on Lake Whitefish Ageing Errors and Northwestern Ontario Baitfish Industry, Technical Report 607. Ottawa: Environment Canada, 1976.

A large (116 pp) collection of data on lacustrine chemistry from Ontario lakes. Also contains data on fish population dynamics in those same lakes. One of the most complete sets of data on these two subjects available. An excellent source of data for analysis or for comparative study.

61. Beck, M., et al. "The Bitter Politics of Acid Rain," Newsweek, April 25, 1983, pp. 36-37.

A journalistic view of the national and international politics involved with discussion of U.S.-Canadian control of acid rain.

62. Becker, R., et al. "Examination of the Acidic Deposition Gradient Across the Lake States Region," Proceedings of a Symposium on Acid Precipitation held in Las Vegas. Washington, DC: American Chemical Society, 1982, pp. 276-277.

An extended abstract of a study that outlines an important regional acidification regime. Very important for scientists.

63. Beigie, C.E. and A.O. hero (eds). Natural Resources in U.S.-Canadian Relations, Vol. II, Boulder, CO: Westview Press, 1980.

An interesting collection of articles by economists, political scientists, and scientists looking at the way in which the U.S. industry relates to Canadian natural resources. The problem of acid rain is not focused on specifically by this book but it is mentioned regularly. Good peripheral reading.

64. Beilke, S. and A.J. Elshout (eds). Acid Deposition. Boston: Reidel Publishing Company, 1983.

Essentially the proceedings of a CEC workshop on atmospheric pollutants held in Berlin in 1982. Articles included range from combustion control to effects to policy analysis. A good international perspective is afforded by the various authors.

65. Beilke, S. "Acid Deposition: The Present Situation in Europe," in Beilke, S. and A.J. Elshout (eds), Acid Deposition. Boston: Reidel Publishing Company, 1983, pp. 3-30.

A general overview of European air pollution problems. Very comprehensive treatment and it speaks specifically to future problems and to present and future research needs. More useful to a scientific or research policy-oriented audience than to a general reader.

66. Bell, H.L. "Effects of Low pH on the Survival and Emergence of Aquatic Insects," Water Resources, Vol. 5, 1971, pp. 313-319.

A good, short report of interest to ecologists and limnologists. Can have effects throughout the aquatic food chain.

67. Bennett, J.H., et al. "Acute Effects of the Combination of Sulphur Dioxide and Nitrogen Dioxide on Plants," Environmental Pollution, Vol. 9, 1975, pp. 127-132.

A basic scientific look at the way in which two of the constituents of acid rain affect plant growth and health. Particularly useful for plant pathologists and agronomists.

68. Benoit, L.F., et al. "Ozone Induced Reduction in Pinus strobus L. Pollen Germination: Potential for Inhibition of Pine Species Reduction by Oxidant Air Pollution," Canadian Journal of Forestry Research, Vol. 13, 1983, pp. 184-187.

Short, technical research report on an atmospheric phenomenon related to acid rain which has a real effect on coniferous evergreens. This kind of species-specific research is important to gaining a comprehensive understanding of combustion effluent effects on the environment. Particularly important because of the focus on the effects on germination, which moves the level of analysis to selective portions of the forest life cycle. Very useful for ecologists as well as foresters.

69. Berner, W., et al. "Past Atmospheric Composition and Climate, Gas Parameters Measured on Ice Cores," Nature, Vol. 276, 1978, pp. 53-55.

Short report of finding in ice cores of layers of atmospheric particles which can give insight into the paleoclimate of certain regions. Not filled with data but useful.

70. Bertine, K.K. and E.D. Goldberg. "Fossil Fuel Combustion and the Major Sedimentary Cycle," Science, Vol. 173, 1971, pp. 233-235.

A non-technical short report on deposition of all types of combustion effluent. Includes topics related to acid rain that may be of interest.

71. Bhumralkar, C.M. (ed). Meteorological Aspects of Acid Rain. Boston: Butterworth Publishers, 1981.

A fine collection of articles on various aspects of the creation and transportation of pollutants from smokestack to acid rain or snow. The articles do require a basic knowledge of meteorogical or climatological principles. The best book to obtain for an in-depth analysis of the meteorology of acid rain.

72. Bick, H. and E.F. Drews. "Self Purification and Silicate Communities in an Acid Milieu," Hydrobiologia, Vol. 42, 1973, pp. 393-402.

Basic scientific research which underlies an understanding of acid rain in aquatic environments. Not specifically focused on acid precipitation.

73. Bigelow, D.S. NADP Instruction Manual. Fort Collins, CO: Natural Resource Ecology Laboratory of Colorado State University, 1982.

Guidelines for establishing a precipitation monitoring site to NADP requirements. Instructions are specific to make and type of equipment and the environmental variables that should be controlled if a collection site is to be suitable. Useful for any field collection study.

74. ______. NADP/NTN Site Selection and Installation. Fort Collins, CO: Natural Resource Ecology Laboratory of Colorado State University, 1984.

Basically an instruction manual for installing a precipitation sampling site to meet NADP and the new National Trends Network standards. Instructions and diagrams are very specific and precise. A short discussion of the background of the NADP/NTN monitoring systems and their technical requirements is also supplied. This is particularly interesting when compared to the site selection criteria outlined in the 1982 instruction manual [item 73].

75. Bjorklund, I., et al. "Mercury in Swedish Lakes--Its Regional Distribution and Causes," Ambio, Vol. 13, no. 2, 1984, pp. 118-121.

Although this article focuses specifically on mercury contamination, it does recognize and analyze the relationship between lacustrine acidity, acidic precipitation, and heavy metal concentrations. A very comprehensive study. Two hundred and twenty typically oligotrophic lakes were in the study. The impact all along the food chain is analyzed.

76. Blake, L.M. "Limiting Acid Ponds in New York," New York Fish and Game Journal, Vol. 28, no. 2, 1981. pp. 208-214.

A report of efforts to restore acidified lakes by artificial liming. The article is written for the general public and does not provide much real technical or scientific insight into the process or the results.

77. _______. "Liming Acid Ponds in New York," in F.M. D'Itri (ed), Acid Precipitation: Effects on Ecological Systems. Ann Arbor, MI: Ann Arbor Scientific Publishing, 1982, pp. 251-260.

A concise report on experiments with agricultural lime applied to five ponds under acid rain attack in upstate New York. Very similar information to article reported above. In addition to the effects, a basic cost analysis for this procedure is provided also.

78. Blanpied, G.D. "Effect of Artificial Rain Water pH and Calcium Concentration on the Calcium and Potassium in Apple Leaves," HortScience, Vol. 14, 1979, pp. 706-708.

Gives a short report of research on fruit tree leaves affected by acid rain. Of use to professional agronomists or orcharders.

79. Bloomfield, J.A., et al. "Atmospheric and Watershed Inputs of Mercury to Cranberry Lake, St. Lawrence County, New York," in T.Y. Toribara, et al. (eds), Polluted Rain. New York: Plenum Press, 1980, pp. 175-210.

Extensive report on research in pollution deposition in a lake environment. Provides data on pH and several other heavy metals in addition to mercury. Gives chemical levels in several fish species as well as lake water chemistry. Also provides data on growth rates of fish related to chemical levels. Very technical but quite useful report.

80. Bolin, B. (ed). Report of the Swedish Preparatory Committee for the U.N. Conference on Human Environment. Stockholm: Norstedt and Soner, 1971.

A clear policy statement of Swedish concern over the international transport of air pollutants and a recommendation for U.N. intervention in this international problem.

81. Bonsang, B., et al. "Sulfate Enrichment in Marine Aerosols Owing to Biogenic Gaseous Sulfur Compounds," Journal of Geophysical Research, Vol. 85, 1980, pp. 7410-7416.

A look at the level of sulfur pollution introduced in the atmosphere over water sources (primarily marshes) by natural cycles of decomposition and effluence.

82. Bormann, F.H. and G.E. Likens. Pattern and Processes in a Forested Ecosystem. New York: Springer-Verlag, 1979.

A good descriptive analysis of the complicated series of cycles and systems active in a forest environment. Mentions acid rain intrusion only briefly but can be used as background reading for an understanding of forest ecology.

83. Bormann, F.H. "The Effects of Air Pollution on the New England Landscape," Ambio, Vol. 11, no. 6, 1982, pp. 338-346.

A regional statement of environmental impact caused by acid rain. Predominately concerned with forest resources (tree kills) and secondarily with water resources. Both phenomena dealt with on a regional rather than a systematic level. A good general regional report.

84. Boutron, C. and R. Delmas. "Historical Record of Global Atmospheric Pollution Revealed in Polar Ice Sheets," Ambio, Vol. 9, 1980, pp. 210-215.

Major ice sheets provide an excellent stratified record of atmospheric deposition for thousands of years. This article provides an overview of climatic evidence, chiefly particulate and chemical deposition, for comparison with modern atmospheric activity.

85. Bowersox, V.C. and R.G. dePena. "Analysis of Precipitation Chemistry at a Central Pennsylvania Site," Journal of Geophysical Research, Vol. 85, 1980, pp. 5614-5620.

A concise research report of data collected at a site in the heart of the eastern U.S. acid rain belt. A good footnote.

86. Bowersox, V.C. and G.H. Stensland. "Seasonal Patterns of Sulfate and Nitrate in Precipitation in the United States," in Proceedings of the 74th Annual Meeting, Air Pollution Control Association, Philadelphia, June 21-26, 1981. Paper no. 81-6.1.

Data which outline the annual variability of acid rain in the eastern U.S. Good source of raw data but weak analysis.

87. Boyle, R.H. and R.A. Boyle. Acid Rain. New York: Schocken Books, 1983.

A good basic book on acid rain for scientist and nonscientist alike. The basic shortcoming is the authors' appeal to the emotional reactions of readers.

88. Bradford, G.R., et al. "Are Sierra Lakes Becoming Acid?" California Agriculture, Vol. 35, no. 5-6, 1981, pp. 7-8.

A short article on reports of research in California lakes. Written for a lay audience, very general and non-technical. Of use to a scientist only because of its regional importance.

89. Bradley, J. M. "West Germany's Great Forests Can No Longer Be Saved," World Environment Report, Vol. 8, No. 1, January, 1982, pp. 1-2.

A very short report of tree kills in Germany caused by acid rain and other pollutants. Gives some general geographic insight into the problem but little else.

90. Bradstreet, T.E., et al. "Vegetation and Associated Environments During the Past 14,000 Years Near Moulton Pond, Maine," Quaternary Research, Vol. 5, 1975, pp. 435-465.

Excellent chronicle of paleoecology in an area that currently has acid rain problems. Good for temporal comparisons.

91. Bradt, P.T. and M.B. Berg. "Preliminary Survey of Pocono Mountain Lakes to Determine Sensitivity to Acid Deposition," Proceedings of the Pennsylvania Academy of Science, Vol. 57, 1983, pp. 190-194.

A short presentation of an analysis of lakes in the Pocono region to determine their geologic and hydrologic characteristics with respect to susceptibility to acidification. A precursor report to the long-term research project [item 92]. A good methodological model for field studies.

92. Bradt, P.T., et al. The Biological and Chemical Impact of Acid Precipitation on Pocono Mountain Lakes. Bethlehem, PA: Department of Biology, Lehigh University, 1984.

A comprehensive (215pp) report on chemical and biological data collection in three lakes in the Pocono Mountains in 1981-83. Data were collected seasonally so the analysis includes subannual phenomena. An outstanding report which includes complete chemical analysis as well as a thorough biotal report in an acidifying region. Good source of data sets as well.

93. Braekke, F.H. (ed). Impact of Acid Precipitation on Forest and Freshwater Ecosystems in Norway, SNSF Project, Report FR#6/76. Oslo: Agricultural Research Council of Norway, 1976.

A comprehensive report of several years of research in Norway on acid rain effects. Similar to reports done in Sweden and useful for comparative analysis with other regional or national reports.

94. _______, et al. "Acidification and Changes Over Time in the Chydoric Cladocera Assemblage of New England Lakes," in G.R. Hendrey (ed). Early Biotic Responses to Advancing Lake Acidification. Boston: Butterworth Publishers, 1984, pp. 85-104.

An examination of lacustrine sediments to determine the chydoric population and speciation changes over time with relation to environmental conditions. Population figures as well as speciation are compared to the pH record.

95. Brandt, C.S. and W.W. Heck. "Effects of Air Pollutants on Vegetation," in A.C. Stern, Air Pollution and Its Effects, 2nd Edition. New York: Academic Press, 1968, pp. 401-443.

A good non-technical article on air pollution effects. Not much emphasis on acid rain but much discussion of the phenomena underlying it and of the mechanisms by which it might enter a plant ecosystem.

96. Brewer, P.F. and A.S. Heagle. "Interactions Between Glomus macrocarpus and Exposure of Soybeans to Ozone or Simulated Acid Rain in the Field," Phytopathology, Vol. 73, 1983, pp. 1035-1040.

A rather technical report on the reactions of microorganisms to acid precipitation. A follow-up to related work done by Shriner and others on soybean plants. Good source of data and analysis.

97. Brezonik, P.L., et al. "Acid Precipitation and Sulfate Deposition in Florida," Science, Vol. 208, 1980, pp. 1027-1029.

A short report on a pollution deposition monitoring program in Florida. Good set of preliminary data. Particularly focused on water resources.

98. Bricker, O.P. (ed). Geological Aspects of Acid Deposition. Boston: Butterworth Publishers, 1981.

A collection of technical papers on the various effects of acid rain on geological resources. This includes the soil, water, and erosion cycles. Very well done and comprehensive but not for non-scientists.

99. Brosset, C. "Air-Borne Acid," Ambio, Vol. 2, no. 1, 1973, pp. 2-9.

A careful look at the meteorological aspects of acid rain production and transportation. Written for ecologists so not as technical as other meteorological treatises.

100. ________. "Characterization of Acidity in Natural Waters," in S. Beilke and A.J. Elshout (eds), Acid Deposition. Boston: Reidel Publishing, 1983, pp. 44-55.

A very technical chemical treatise of the ionic reactions in water under the effects of acid rain. Makes the point that, to fully understand the chemical dynamics of water, a measurement of pH is simply insufficient.

101. Brouzes, R.J.P., et al. The Link Between pH of Natural Waters and the Mercury Content of Fish. Senneville, Quebec: Domtar Research Center, 1977.

A thorough report of research on fish populations in Canada. Establishes a positive correlation between lake and stream acidity and mercury contamination in fish. Very useful sets of data and analysis for comparative study.

102. Brydges, T. "Aquatic Effects of Acid Deposition--Water Quality," Proceedings of the APCA Conference on Acid Deposition, April 7-8, 1981. Montreal: APCA, 1981, pp. 45-56.

An analysis of water chemistry in natural systems under attack by acid rain. Particularly useful because of the information on subsurface dynamics.

103. Bubenick, D.V. and F.A. Record. "Acid Rain--An Overview of the Problem," Environmental Progress, Vol. 2, no. 1, 1983, pp. 15-31.

A general article for a non-scientific audience. Does touch on all aspects of acid rain impact but does not go into detail on any one. Also deals with policy and international politics to some degree.

104. Budiansky, T. "Acid Rain and the Missing Link," Environmental Science and Technology, Vol. 14, no. 10, 1980, pp. 1172-1173.

A very surficial comment on acid rain research and publication. It does serve to point to several areas of research need with regard to acid rain modeling. No new data or analysis forthcoming.

105. _______. "Understanding Acid Rain," Environmental Science and Technology, Vol. 15, no. 6, 1981, pp. 623-624.

A journalistic article treating the phenomenon very generally. More for a lay audience.

106. Burns, D.A., et al. "Acidification of Surface Waters in Two Areas of the Eastern United States," Water, Air, and Soil Pollution, Vol. 16, 1981, pp. 277-285.

Thorough regional study of stream and lake acidification. Begins to delineate between precipitation and land-use induced effects.

107. Burton, C.S. and G.M. Hidy. Regional Air Pollution Study Program Objectives and Plans, EPA 650/3-75-009. Research

Triangle Park, NC: U.S. Environmental Protection Agency, 1974.

Proposes development of several types of pollution-monitoring programs across the U.S., including precipitation pH network which ultimately became reality. Good source of information for policy scientists.

108. Burton, T.M., et al. "Effects of Acid Rain on Michigan Streams," Michigan Riparian, November, 1981, pp. 6-7.

Short but informative regional update on stream quality monitoring. Written for general informed audience.

109. ________. "The Effects of Acidification on Stream Ecosystems," in F.M. D'Itri (ed), Acid Precipitation Effects on Ecological Systems. Ann Arbor, MI: Ann Arbor Science Publishers, 1982, pp. 209-235.

A particularly detailed overview of the chemistry and ecology of streams under acid rain attack. Most useful for the contrasts it makes with ecology in lacustrine environments.

110. Bush, D.M. The Potential Impact of Acid Precipitation on Wisconsin's Fisheries, Fish Management Report 105. Madison, WI: Wisconsin Department of Natural Resources, 1980.

An applied analysis of acid rain impact based on information from Wisconsin but applicable to all regions. Good base-line data here for comparative study.

111. Cadle, R.D. "A Comparison of Volcanic With Other Fluxes of Atmospheric Trace Gas Constituents," Review of Geophysical and Space Physics, Vol. 18, 1980, pp. 746-752.

Surficial analysis of the effect of volcanic activity on atmospheric chemistry.

112. Calvert, J.G., (ed). SO_2, NO and NO_2 Oxidation Mechanisms. Boston: Butterworth Publishers, 1982.

A comprehensive compilation of articles on atmospheric chemistry and physics that underlies the production of acid rain. Good, solid scientific articles not for the non-scientific audience but necessary for a scientific understanding of acid precipitation.

113. Campbell, S.A. "Standardization for Precipitation Studies," in D.H. Pack and A.A. Shepherd, (eds), Proceedings: Advisory Workshop on Methods for Comparing Precipitation

Chemistry Data. Washington,DC: Utility Acid Precipitation Study Program, 1982, pp. 4-57--4-66.

A call for more compatibility in data gathering by groups that monitor precipitation chemistry. Some of the recommendations are useful but the tone of the article is very heavily pro-industry; it is an attempt to discredit much of the research on acid rain that points at electric generating plants as culprits.

114. Canfield, D.E., Jr. "Sensitivity of Florida Lakes to Acidic Precipitation," Water Resources Research, Vol. 19, no. 3, 1983, pp. 23-29.

A thorough report on the geological characteristics and hydrologic properties of lakes in Florida that affect their susceptibility to acidification. In general, these lakes contain a large amount of natural sulfur but they are often buffered from acidification by limestone bedrock. A good source for a comparative study with other regions.

115. Carlson, R.W., et al. "Physiological Effects, Wind Reentrainment, and Rain Wash of Pb Aerosol Particulate on Plant Leaves," Environmental Science and Technology, vol. 12, 1976, pp. 1139-1142.

A very technical article on the effects of lead deposition on plant leaves and health. Many of the same processes and effects are involved with and related to acid rain. A good background article.

116. Carlson, R.W. "Reduction in the Photosynthetic Rate of Acer, Quercus and Fraxinus Species Caused by Sulphur Dioxide and Ozone," Environmental Pollution, Vol. 18, 1979, pp. 159-170.

A rather general article documenting reduced rates of photosynthesis in selected trees in the presence of SO_2 and ozone. Useful as part of the whole range of research on forest effects.

117. Carmichael, G.R. and L.K. Peters. "Numerical Simulation of the Regional Transport of SO_2 and Sulfate in the Eastern United States," Proceedings of the 4th Symposium on Turbulence, Diffusion, and Air Pollution. Boston: American Meteorlogical Society, 1979, pp. 337-344.

Very technical and comprehensive examination of atmospheric transport models used to predict acid rain downwind of central U.S. pollution sites.

118. Carroll, J.E. Acid Rain: An Issue in Canadian-American Relations. Washington,DC: National Planning Association, 1982.

A very good synopsis of the political problems created by acid rain. Some basic policy information contained within and an excellent analysis of political as well as scientific considerations involved in the development of these policies. Necessary reading for all policy students and quite useful for scientists as well.

119. ________. Environmental Diplomacy: An Examination and a Perspective of Canadian-United States Transboundary Environmental Relations. Ann Arbor, MI: University of Michigan Press, 1983.

A comprehensive presentation, discussion, and analysis of the development of international environmental policy between the U.S. and Canada. Documents the political as well as regulatory input in policy development and gives considerable discussion to policies, particularly acid rain oriented ones, that have not been developed and are conspicuous by their absence. Must reading for policy students.

120. Carroll, P. "Differences in the Environmental Regulatory Climate of Canada and the United States," Canadian Water Resources Journal, Vol. 14, Fall, 1979, pp. 13-19.

A good comparison of the attitudes of both countries toward environmental policy. The differences created by being a receiver rather than a sender are quite clearly outlined. One of the critical articles if one is to follow the development of a joint U.S.-Canadian policy on acid rain.

121. Carter, J. The President's Environmental Program 1979. Washington, DC: The President's Council on Environmental Quality, 1979.

Basically a compendium of facts gleaned from Federal agency reports. Acid rain is just one of the environmental problems dealt with. No new information but an attractive presentation.

122. Carter, L.J. "Uncontrolled SO_2 Emissions Bring Acid Rain," Science, Vol. 204, June, 1979, pp. 1179-1182.

Recognizes the host of problems created by the EPA tall stacks policy and the lack of upper air pollution policy in the U.S. Gives a general outline of the scientific principles associated with acid rain.

123. Cavender, J.H., et al. Interstate Surveillance: Measurements of Air Pollution Using Static Monitor, Report No. APTD 777. Research Triangle Park, NC: U.S. Environmental Protection Agency, 1971.

An early source of national and regional atmospheric chemistry data developed in experimental data collection system. Very good source for temporal or geographic comparisons.

124. ______. Nationwide Air Pollutant Emission Trends, 1940-1970. Research Triangle Park, NC: U.S. Environmental Protection Agency Laboratory, 1973.

Comprehensive compilation of raw data taken from a national network of precipitation monitoring programs. Useful only if one is ready to plow through a large computerized "dump" of information. Very useful as a data set for large scale analysis.

125. Cawse, P.A. A Survey of Atmospheric Trace Elements in the United Kingdom, A.E.R.E. Hawwell Report No. R-7669. London: HMSO, 1974.

Good collection of raw data on atmospheric pollution in the British Isles. As a major source area for acid rain for western Europe, the air quality of the U.K. is very important.

126. Chagnon, S.A. Precipitation Scavenging of Lake Michigan Basin, Bulletin No. 52. Urbana, IL: Illinois State Water Survey, 1968.

A good technical report on specific meteorological events in the Great Lakes region. Filled with usable data for comparative study.

127. Chamberlain, A.C. "Dry Deposition of Sulfur Dioxide," in D.S. Shriner, et al. (eds), Atmospheric Sulfur Deposition. Ann Arbor, MI: Ann Arbor Science Publishers, 1980, pp. 185-197.

Good scientific introduction to dry deposition processes, a phenomenon of much less importance than wet deposition but of interest to students of depositional processes. Not recommended for the casual non-scientific reader.

128. Chameides, W.L. and D.D. Davis. "Chemistry in the Troposphere," Chemical and Engineering News, Vol. 60, No. 40, 1982, pp. 38-52.

A look at overall atmospheric chemistry, some of which is the basis for the acid precipitation phenomenon.

129. Chamile, W.H., et al. "Airborne Sulfur Dioxide to Sulfate Oxidation Studies of the INCO 381m Chimney Plume," Atmospheric Environment, Vol. 14, 1980, pp. 1159-1170.

A very technical analysis of the aerosol chemistry of effluent from the world's tallest smokestack. A good analysis of the single largest air pollution site in North America.

130. Chan, S.G., et al. "The Importance of Soot Particles and Nitrous Acid in Oxidizing SO_2 in Atmospheric Aqueous

Droplets," Atmospheric Environment, Vol. 15, 1981, pp. 1287-1292.

A very technical report on basic atmospheric chemistry. Particularly useful for an analysis of the effects scrubbers are having on acidification.

131. Chan, W.H., et al. "Precipitation Scavenging and Dry Deposition of Pollutants from the INCO Nickel Smelter in Sudbury," Proceedings of the Fourth International Conference on Precipitation Scavening, Dry Deposition, and Resuspension. Santa Monica, CA: Univ. of California, 1983, pp. 56-77.

An excellent evaluation of the processes involved with localized acid rain production near a major sulfur polluter. Basic atmospheric physics and chemistry.

132. Charlson, R.J. and H. Rodhe. "Factors Controlling the Acidity of Natural Rainwater," Nature, Vol. 295, 1982, pp. 683-685.

A very general overview of the natural and anthropogenic acidification of precipitation. Looks at the atmosphere as a chemical sink and illustrates how it accepts and rejects effluent.

133. Cheney, J.L. and J.B. Homolya. Workshop Proceeding on Primary Sulfate Emissions From the Combustion of Fossil Fuels, EPA-600/9-78-020a. Research Triangle Park, NC: U.S. Environmental Protection Agency, 1978.

Report of discussions of basic research into the overall national problem of combustion control. More for policy analysts than for scientists.

134. Chevone, B.I. and Y.S. Yang. "Precipitation Chemistry in the Southern Appalachian Mountains of Virginia," Phytopathology, Vol. 72, no. 6, 1982, p. 706.

A short report of data collected in a relatively pristine region in the eastern United States. Little analysis, but a usable source of data and information about acid rain research on a regional basis.

135. Chevone, B.I., et al. "Acidic Precipitation and Ozone Effects on Growth of Loblolly and Shortleaf Pine Seedlings," Phytopathology, Vol. 74, no. 6, 1984, p. 756.

An abstract of a verbal report on simulated acid rain applied to very young trees. These two species of important economic forest crops reacted negatively to acid rain applications with significant root growth reductions. There was no corresponding overall growth retardation but the implications for future development with reduced root networks is quite negative. Useful for forest ecologists.

136. Chung, Y.S. "The Distribution of Atmospheric Sulfates in Canada and Its Relationship to Long-range Transport of Pollutants," Atmospheric Environment, Vol. 12, 1978, pp. 1471-1480.

A solid technical report of effluent sources and the effects this can have on redistribution and deposition of acid rain.

137. Claiborne, R. Climate, Man and History: An Irreverent View of the Human Environment. New York: W.W. Norton and Co., 1970.

General reading for scientist and non-scientist alike. Looks at the many ways in which human cultural evolution has affected and been affected by the climate.

138. Clair, T.A. and P.H. Whitfield. "Trends in pH, Calcium, and Sulfate of Rivers in Atlantic Canada," Limnological Oceanography, Vol. 28, no. 1, 1983, pp. 160-165.

A rather technical but generally readable report on field data collected in the maritime provinces of Canada. Important not only because of the quality of the study but because of the sensitivity of the region.

139. Clark, H.L., et al. "Acid Rain in Venezuelan Amazon," in Tropical Ecology and Development, Kuala Lumpur, Malaysia: International Society of Tropical Ecology, 1982, p. 683.

A very short research note on pollution in a pristine environment. Although there is very little data presented, this citation illustrates the truly global nature of the acid rain phenomenon.

140. Clark, K. and K. Fischer. Acid Precipitation and Wildlife, Wildlife Toxicology Series No. 43. Ottawa: Canadian Wildlife Service, 1981.

A good, general overview of the effects of acid rain on forest and aquatic life systems. Written by scientists for a public audience.

141. Clark, W.W., et al. "Measurements of the Photochemical Production of Aerosols in Ambient Air Near a Freeway for a Range of SO_2 Concentrations," Atmospheric Environment, Vol. 10, 1976, pp. 637-644.

A clear examination and report of data collection on one specialized segment of atmospheric chemistry that can affect acidification on a local level. Very well presented but very specialized information.

142. Clement, P. and J. Vandour. "Observations on the pH of Melting Snow in the Southern French Alps," Arctic and Alpine Environments, Vol. 34, 1967, pp. 205-213.

Report on the micro and macro processes of snow melt in a European setting. Good for a basic scientific understanding of the way in which acid moves through the snow cycle.

143. Cocks, A.T. and W.J. McElroy. "Modeling Studies of the Concurrent Growth and Neutralization of Sulfuric Acid Aerosols Under Conditions in the Human Airways," Environmental Research, Vol. 35, no. 1, 1984, pp. 79-86.

Primarily oriented toward medical research. Examines gas phase reactions that can affect human respiratory health. Looks at events during famous London acid fogs and extrapolates to include inferences for general acidic atmospheric conditions. Useful especially in light of the dearth of data on the health implication of acidification.

144. Cogbill, C.V. and G.E. Likens. "Acid Precipitation in the Northeastern United States," Water Resources Research, Vol. 10, 1974, pp. 1133-1137.

An overview of the first region in the U.S. that was affected by acid rain. Figures are dated and very general but can provide continuity for a longer study.

145. Cogbill, C.V. "The History and Character of Acid Precipitation in Eastern North America," Water, Air, and Soil Pollution, Vol. 6, 1976, pp. 407-413.

A relatively non-technical developmental analysis of acid rain with a regional focus. Good basic information article with some regional analysis and synthesis of a wide range of research not only by the author but by others.

146. ________. "The Effect of Acid Precipitation on Tree Growth in Eastern North America," in L.S. Dochinger and T.S. Seliga (eds), Proceedings of the First International Symposium on Acid Precipitation and the Forest Ecosystem, General Technical Report NE-23. Columbus, OH: U.S. Forest Service, 1976, pp. 1027-1032.

Primarily predictive discussion based on models of forest ecosystems rather than on field data, due to scarcity of data at that date. Good general reference.

147. Cohen, C.J., et al. Effects of Simulated Sulfuric Acid Rain on Crop Plants, Special Report 619 of the Agricultural Experiment Station, Oregon State University, May, 1981.

Very technical and important report of the ways in which acid rain does and doesn't affect certain crops. The basis for economic and agricultural predictions of regional and national effects on crop production.

148. ________. Effects of Simulated Sulfuric and Sulfuric-Nitric Acid Rain on Crop Plants, Special Report 670 of the Agricultural Experiment Station, Oregon State University, November, 1982.

A comprehensive listing of test results on agricultural plants that were subjected to simulated acid rain. Useful for all levels of interest yet scientifically very significant. Data are in chart form but narrative helps to interpret and bring them together.

149. Cole, A.F.W. and M. Katz. "Summer Ozone Concentrations in Southern Ontario in Relation to Photochemical Aspects and Vegetation Damage," Journal of the Air Pollution Control Association, Vol. 16, 1966, pp. 201-206.

Written well before acid rain became a national problem, this article documents seasonal fluctuations in combustion effluent and correlates the rises in these chemicals with vegetative damage. Vegetative study is general and surficial but the atmospheric chemical models are first rate.

150. Cole, D.W. and D.W. Johnson. "Atmospheric Sulfate Additions and Cation Leaching in a Douglas Fir Ecosystem," Water Resources Research, Vol. 13, 1977, pp. 313-317.

An examination of movement of one important chemical segment of acid rain through a forest system. The implications for soil utility apply to all forest systems. Very useful article.

151. Cole, P.W. (ed). Acid Rain: A Transjurisdictional Problem in Search of Solution. Buffalo, NY: State University of New York, 1982.

One of several studies of the political and legal problems created by acid rain. A collection of articles which deals with most policy and legal aspects of the U.S.-Canada acid rain negotiations.

152. Collins, R. "Acid Rain: Scourge From the Skies," Reader's Digest, January, 1981, pp. 109-113.

General public article that provides an overview of acid rain problem.

153. Collison, R.C. and J.E. Mensching. Lysimeter Investigations: Composition of Rain Water at Geneva, NY, for a 10-Year Period, Technical Bulletin No. 193 of the New York Experiment Station, Geneva, N.Y., 1932.

Simply a compilation of very early data on precipitation chemistry in an area that is now suffering heavily from acid rain.

154. Commins, B.T. "Determination of Particulate Acid in Town Air," Analyst, Vol. 88, 1963, pp. 364-367.

A non-technical examination of methods used to measure atmospheric acidification in pre-tall stacks situations of intense local pollution.

155. Connell, D.W. and G.J. Miller. Chemistry and Ecotoxicology of Pollution. New York: John Wiley and Sons, 1984.

A very technical text and very comprehensive in its coverage of the subject. Only one chapter specifically focused on acid rain but some excellent scientific discussions of processes and effects that are related to acid rain.

156. Conservation Foundation. Acid Rain: A Major Threat to the Ecosystem. Washington, DC: The Conservation Foundation, 1982.

A very good look at the environmental and public policy problems associated with acid rain in North America. The Foundation does share its pro-environmental bias with the reader.

157. _______. Will Congress Swallow an Anti-Acid Bill?, Washington, DC: The Conservation Foundation, 1983.

An environmentalist's analysis of the politics of creating a U.S. policy on acid rain. National and international implications are presented.

158. Constantinidou, H.A. and T.T. Kozlowski. "Effects of Sulphur Dioxide and Ozone on Ulmus americana Seedlings: Visible Injury and Growth," Canadian Journal of Botany, Vol. 57, 1979, pp. 170-175.

A thorough research report on controlled-environment tests of plants exposed to SO_2 and ozone, both singularly and in combination. Results have implications for all plants exposed to combustion effluent.

159. Cooley, J.M. "Aquatic Effects: Water Quality and Fisheries," Proceedings of the APCA Conference on Acid Deposition, April 7-8, 1981. Montreal: APCA, 1981, pp. 29-33.

A concise overview of general effects of acid rain on water quality as it affects commercially important or sport fish. Useful for general scientists.

160. Corbett, E.S. and J.A. Lynch. "Rapid Fluctuations in

Streamflow pH and Associated Water Quality Parameters During a Stormflow Event," Proceedings of the International Symposium on Hydrometeorology. Denver, CO: University of Denver, 1983, pp. 461-464.

A rather technical report on stream pH during a storm situation. The data provided here are especially critical because of the special conditions that exist during precipitation events. The maxima and minima established during these events may be controlling factors in overall stream health. Useful background information for all scientists.

161. Correll, D.L. and D. Ford. "Comparison of Precipitation and Land Runoff as Sources of Estuarine Nitrogen," Estuarine, Coastal and Shelf Science, Vol. 15, 1981, pp. 45-46.

A short but complete article on the levels of nitrogen pollution in selected U.S. estuaries. An analysis of the sources of this pollution is attempted to delineate acid rain input from runoff, generally thought to come from agricultural regions. The results are inexact and incomplete but quite useful and the methodological implications are of interest.

162. Costle, R. "New Source Performance Standards for Coal-fired Power Plants," Journal of Air Pollution Construction Association, Vol. 29, July, 1979, p. 690.

An outline of EPA regulations that pertain to coal combustion and the new construction needed to help meet these standards. Policy as it is implemented.

163. Council on Environmental Quality. Environmental Quality 1981. Washington, DC: U.S. Government Printing Office, 1982.

A compendium of data from governmental agencies that work with natural resources and the environment. No new information but a bureaucratically sound compilation of material into a usable document.

164. Cowell, D.W., et al. The Development of an Ecological Sensitivity Rating for Acid Precipitation Impact Assessment, Working Paper No. 10. Burlington, Ontario: Environment Canada, 1981.

A policy paper that approaches the idea of creating an environmental impact assessment program tailored specifically to the effects of acid rain. As a policy document it has significant amounts of science incorporated.

165. Cowling, E.B. "Effects of Acid Precipitation and Atmospheric Deposition on Terrestrial Vegetation," in J.N. Galloway, et al., (eds), A National Program for Assessing the Problem of

Atmospheric Deposition (Acid Rain), National Atmospheric Deposition Program report to the U.S. Council on Environmental Quality, NC-141. Washington, DC: U.S. Government Printing Office, 1978, pp. 46-63.

A very comprehensive overview of the known and expected effects of acid rain on vegetation. This draws from European research heavily and, because of its relatively early date, it offers implications rather than the explicit information that is now available.

166. _______ and L.S. Dochinger. "Effects of Acidic Precipitation on Health and the Productivity of Forests," in P.R. Miller (ed), Effects of Air Pollutants on Mediterranean and Temperate Forest Ecosystems, General Technical Report PSW-43. Washington, DC: U.S. Forest Service, 1980, pp. 166-173.

A very comprehensive overview of forest ecosystem reaction to acid precipitation. Written by scientists but clearly enough for a non-scientific audience.

167. _______ and R.A. Linthurst. "The Acid Precipitation Phenomenon and Its Ecological Consequences," Bioscience, Vol. 31, No. 9, 1981, pp. 649-654.

A general overview of ecology and acid rain. It is comprehensive in touching subjects but rather shallow in treating specific subjects, especially for a scientific journal article.

168. _______. An Historical Resume of Progress in Scientific and Public Understanding of Acid Precipitation and Its Biological Consequences. As, Norway: SNSF Project, 1981.

A very good look at the development of public and scientific awareness of acid rain and its impact. This provides a scientist's view of public policy evolution.

169. _______. "A Status Report on Acid Precipitation and Its Biological Consequences as of April, 1981," in F.M. D'Itri (ed.), Acid Precipitation: Effects on Ecological Systems. Ann Arbor, MI: Ann Arbor Science Publishers, 1982, pp. 1-20.

A very comprehensive overview of the acid rain phenomenon and of research into the effects of acid rain on the environment. Recommended for general scientific reading and as a starting point for a potential acid rain researcher.

170. _______. "Acid Precipitation in Historical Perspective," Environmental Science and Technology, Vol. 16, no. 2, 1982, pp. 110-123.

History here being primarily the very recent, as is awareness of acid rain. A well-written documentation by one of the international leaders in acid rain research.

171. _______. "Acid Rain and Atmospheric Deposition: Causes, Consequences and Research Needs," Proceedings of the annual meeting of the Technical Association of the Pulp and Paper Industry. Washington, DC: TAPPI, 1983, pp. 119-123.

A general overview of the process and problem of acid rain production in the U.S. written for a general audience. The portion on research needs is particularly interesting and insightful.

172. Coyne, M. and N. Smith. For Crying Out Cloud: A Study of Acid Rain. Minneapolis, MN: Tasa Publishing Company, 1981.

A book for the general public written from an "environmentalist" perspective. Of use to policy analysts as a gauge of public concern and interest.

173. Cragin, J.H., et al. The Chemistry of 700 Years of Precipitation at Dye-3 Greenland. Hanover, NH: U.S. Army Cold Regions Research and Engineering Laboratory, 1975.

A data collection that is very useful to students of global or regional climatic fluctuation. The Cold Regions lab is an international standard in scientific research.

174. Craker, L.E. and D. Bernstein. "Buffering of Acid Rain by Leaf Tissue of Selected Crop Plants," Environmental Pollution, Vol. 36, no. 4, 1984, pp. 375-381.

Leaves placed in simulated sulphuric acid rain revealed that the buffering ability of selected crop plants (wheat, soybean, corn) differs significantly and may affect their overall resistance to acid rain. Good source of basic research on plant effects.

175. Crocker, T.D. and B.A. Forster. "Decision Problems in the Control of Acid Precipitation: Nonconvexities and Irreversibilities," Journal of the Air Pollution Control Association, Vol. 31, no. 1, 1981, pp. 31-37.

An analysis of the scientific and policy development problems involved with regulation of acid rain. A good presentation which intertwines economic, scientific, and political processes.

176. Crocker, T.D. "Pollution Induced Damages to Managed Ecosystems: On Making Economic Assessments," in J.S.

Jacobson and A.A. Miller (eds), Effects of Air Pollution on Farm Commodities. Arlington, VA: Isaak Walton League, 1982, pp. 103-124.

A relatively non-technical guide to the way in which economists can and do make estimations of pollution damage to agriculture in the United States. Very useful to economists, policy scientists, and general agricultural analysts.

177. _______, ed. Economic Perspectives on Acid Deposition Control. Boston: Butterworth Publishers, 1984.

A very technical compilation of articles looking at various aspects of economic analysis related to acid rain effects and regulation. Not recommended for non-economists or even non-quantitative economists.

178. _______. "Scientific Truths and Policy Truths in Acid Deposition Research," in T.D. Crocker (ed), Economic Perspectives on Acid Deposition Control. Boston: Butterworth Publishers, 1984, pp. 65-80.

A statistical evaluation of the two contrary views on the causes of acid rain in the U.S. as held by the pro-electric utility and the anti-electric utility forces. A very technical look at the economic factors involved with regulatory development and the level of truth associated with scientific information provided to policy-makers.

179. Cronan, C.S., et al. "Forest Floor Leaching: Contributions From Mineral, Organic and Carbonic Acids in New Hampshire Subalpine Forests," Science, Vol. 200, 1978, pp. 309-311.

A thorough report on the movements of acids and acid-related minerals through a forest soil. The impact of acids from natural and acid rain sources is clearly delineated. Good source of analysis but not much data presented.

180. Cronan, C.S. and C.L. Schofield. "Aluminum Leaching Response to Acid Precipitation," Science, Vol. 204, 1979, pp. 304-306.

Short research report on ionic activity in forest soils under stress from acid rain.

181. Cronan, C.S. "Consequences of Sulfuric Acid Inputs to a Forest Soil," in D.S. Shriner, et al., (eds), Atmospheric Sulfur Deposition. Ann Arbor, MI: Ann Arbor Scientific Publishing, 1980, pp. 334-343.

An in-depth analysis of soil chemistry as it is affected by one segment of acid rain. This is now a standard reference on this subject.

182. Cross, R.F. "Smuggled in by the Wind," The New York Conservationist, Vol. 33, no. 5, 1979, pp. 36-38.

A journalistic article for a general public audience. Useful for a beginning interest.

183. Crowther, D.J. Precipitation Chemistry at a Central New York Site, M.S. Thesis. Ithaca, NY: Cornell University Library, 1984.

A good report of data collected in an important acid rain-impacted region. Very little analysis provided or attempted but a useful source of precipitation data.

184. Currie, D.P. "Direct Federal Regulation of Stationary Sources Under the Clean Air Act," University of Pennsylvania Law Review, Vol. 128, no. 6, 1980, pp. 1389-1470.

Rather technical evaluation of the policy behind regulation of single source polluters by the U.S. government under the new provisions of the Clean Air Act. Good for policy studies.

185. Curtis, C. Before the Rainbow: What We Know About Acid Rain. Washington, DC: Edison Electric Institute, 1980.

An overview of the acid rain phenomenon from the viewpoint of an electric utility lobbying organization. Some useful data presented but the analysis is more advocacy than pure science.

186. Dams, R. and J. DeJonge. "Chemical Composition of Swiss Aerosols from the Jungfraujoch," Atmospheric Environment, Vol. 10, 1976, pp. 1079-1084.

Basic scientific data presentation from a major effluent site in western Europe. Just at the beginning of European awareness of the extent and intensity of acid rain effects.

187. Dana, M.T., et al. "Rain Scavenging of SO_2 and Sulfate from Power Plant Plumes," Journal of Geophysical Research, Vol. 80, 1975, pp. 4119-4129.

A very scientific analysis of the atmospheric physics involved with uptake of sulfur compounds from effluent sites. Not recommended for non-scientists.

188. Dana, M.T. and D.W. Glover. Precipitation Scavenging of Power Plant Effluents: Rainwater Concentrations of Sulfur and Nitrogen Compounds and Evaluation of Rain Samples Desorption of SO_2, BNWL-1950. Washington, DC: U.S. Atomic Energy Commission, 1975.

Solid technical report written for bureaucratic consumption but useful as both a scientific data source and as a document to support policy analysis.

189. Dana, M.T. and J. Hlaes. "Measurement of Precipitation Chemistry," MAP3S Progress Report. Washington, DC: U.S. Department of Energy, 1979, p. 72.

A short note on detailing the development of a national precipitation monitoring program.

190. Dana, M.T. Distribution of Contaminants. Columbus, OH: Batelle Columbus Laboratories, 1980.

Short (79pp) report to the electric utilities on the geographic patterning of atmospheric chemicals thought to come from combustion sites. Good data source.

191. _______, et al. "Wintertime Precipitation Chemistry in North Georgia," Proceedings of the Symposium on Acid Rain held in Las Vegas, Nevada. Washington, DC: American Chemical Society, 1982, pp. 115-122.

Very useful regional data on precipitation chemistry. The region studied is important particularly because of its agriculture and forest resources.

192. Dasgupta, P.K. "Discussion of the Importance of Atmospheric Ozone and Hydrogen Peroxide in Oxidizing Sulphur Dioxide in Cloud and Rainwater," Atmospheric Environment, Vol. 14, 1980, pp. 272-274.

A relatively shallow research note looking at hypothetical reactions in atmospheric chemistry that underlie the production of acid rain.

193. Davidson, C.I. et al. "The Influence of Surface Structure on Predicted Particle Dry Deposition to Natural Grass Canopies," Water, Air, and Soil Pollution, Vol. 18, 1981, pp. 25-43.

Very technical presentation of the ways in which particles of aerosol pollution will or will not adhere to natural vegetative surfaces.

194. _______. "Wet and Dry Deposition of Trace Elements into the Greenland Ice Sheet," Atmospheric Environment, Vol. 15, no. 19, pp. 1-9.

A non-analytical report of field study of the record of atmospheric deposition in one of the oldest ice sheets extant. Good paleo-data that can be of use in establishing past trends in atmospheric chemistry.

195. Davies, J.C. and B.S. Davies. The Politics of Pollution. Indianapolis, IN: Pegasus Press, 1975.

Primarily a book on policy issues related to pollution control. Includes sections on air pollution legislation, the courts, and the Clean Air Act, and federal enforcement of air pollution control regulations. Very useful as an introduction to federal regulatory interest and action in the area of air pollution control.

196. Davies, T.D., et al. "Preferential Elution of Strong Acids From a Norwegian Ice Cap," Nature, Vol. 300, 1982, pp. 161-163.

A general, short presentation on how acids are stored and released by ice and snow.

197. Davies, T.D. and K.W. Nicholson. "Dry Deposition Velocities of Aerosol Sulphate in Rural Eastern England," in H.W. Georgii and J. Pankrath (eds), Deposition of Atmospheric Pollutants. Boston: Reidel Publishing, 1982, pp. 31-42.

A report of the results of x-ray examination of deposited material in rural England. Reports a seasonal variation in deposition rates and variation with atmospheric stability and other climatic conditions. Of use to climatologists, meteorologists and general ecologists.

198. Davis, R.B. and R. Berge. "Atmospheric Deposition in Norway During the Last 300 Years as Recorded in SNSF Lake Sediments," in D. Drablos and A. Tollan (eds), Proceedings of the International Conference on Ecological Impact of Acid Precipitation. Sandefjord, Norway: SNSF Project, 1980, pp. 270-271.

A synopsis of an ongoing research into the use of varve records to infer paleoatmospheric chemistry. Very useful report for scientists and non-scientists alike.

199. Davis, W.E. and L.L. Wendell. "Some Effects of Isentropic Vertical Motion Simulation in a Regional-scale Quasi-Lagrangian Air Quality Model," Proceedings of the Third Symposium on Atmospheric Turbulence, Diffusion, and Air Quality. Raleigh, NC: American Meteorological Society, 1976, pp. 403-406.

Very technical specialized information basic to development of infusion and transportation of effluent into and by the atmosphere. A look at how chemicals get into the upper atmosphere; not for non-scientists.

200. Dawson, G.A. "Atmospheric Ammonia From Undisturbed Land," Journal of Geophysical Research, Vol. 82, 1977, pp. 3125-3133.

Good examination of a natural process of acidification, useful for comparison with anthropogenic sourcing.

201. ________. "Ionic Composition of Rain During Sixteen Convective Showers," Atmospheric Environment, Vol. 12, 1978, pp. 1991-1999.

Basic scientific data detailing specific precipitation chemistry for specific events. Good data for scientific readers.

202. Daye, P.G. and E.T. Garside. "Lethal Levels of pH for Brook Trout," Canadian Journal of Zoology, vol. 53, 1975, pp. 639-641.

Concise research report with usable data presented for limnologists and ecologists. Establishes toxicity baselines for an important species of sport fish.

203. Deland, M.R. "Acid Rain," Environmental Science and Technology, Vol. 14, No. 6, 1980, p. 657.

A general overview of the state of knowledge about this complex subject in abbreviated form. Really a synopsis of the state of knowledge on the subject at that time.

204. Delmas, R. and C. Boutron. "Sulfate in Antarctic Snow: Spatio-temporal Distribution," Atmospheric Environment, Vol. 12, 1978, pp. 723-728.

An examination of the atmospheric pollution record provided in packed snow. More concerned with the physical and chemical processes involved than with the information the record provides.

205. Delmas, R. and A. Aristariain. "Recent Evolution of Strong Acidity of Snow at Mt. Blanc," in Proceedings of the 13th Annual International Colloquium on Atmospheric Pollution, Vol. 1. Amsterdam: Elsevier, 1979, pp. 233-237.

Short report of acid deposition findings on a major tourist site in Europe. Significant economic and social consequences are implied because of the importance of tourism in this region.

206. Delmas, R., et al. "Emissions and Concentrations of Hydrogen Sulfide in the Air of the Tropical Forest of the Ivory Coast and of Temperate Regions in France," Journal of Geophysical Research, Vol. 85, 1980, pp. 4468-4474.

An excellent scientific comparison of the atmospheric chemistry of one of the most pristine regions in the world with one in France which is under attack by acid rain. The comparison is very useful for delineating "natural" phenomena from anthropogenic.

207. Delmas, R. and C. Boutron. "Are the Past Variations of the Stratospheric Sulfate Burden Recorded in Central Antarctic Snow and Ice Layers?" Journal of Geophysical Research, Vol. 85, 1980, pp. 5645-5649.

Background information on evidence and processes that infer the title of this article. This is a research proposal and, as such, provides targeted data on the phenomenon in general.

208. Delmas, R.J. "Antarctic Sulphate Budget," Nature, Vol. 299, 1982, pp. 677-678.

A short report of continuing antarctic research into cold region air pollution. Very useful as a datum point from which to compare later chemical readings from such regions.

209. DelMonte, M., et al. "Airborne Carbon Particles and Marble Deterioration," Atmospheric Environment, Vol. 15, 1981, pp. 645-652.

An assessment of the effects of one component of acid rain on building materials. Not of interest to "environmental scientists" per se.

210. Dennison, R., et al. "The Effects of Acid Rain on Nitrogen Fixation in Western Washington Coniferous Forests," Proceedings of the First International Symposium on Acid Precipitation and the Forest Ecosystem, General Technical Report NE-23. Columbus, OH: U.S. Forest Service, 1976, pp. 933-950.

A basic research article on field tests of subsurface forest ecology. The negative implications for nitrogen-fixing bacteria have relevance not only to forests but to agriculture.

211. Devitt, T.W. and T.C. Ponder, Jr. "Status of Sulfur Dioxide Removal Systems for the Electric Utility Industry," in K.E. Noll and W.T. Daibs (eds), Power Generation: Air Pollution Monitoring and Control. Ann Arbor, MI: Ann Arbor Science Publishers, 1976, pp. 49-64.

An initial evaluation of the flue gas desulfurization (FGD) technology in the U.S. as of this date. Provides specific data on twenty-one sites. Also evaluates cost of this and of more advanced technology. Good reference for pollution control scientists and also useful for policy students.

212. Dewalle, D.R., et al. "Acid Snowpack Chemistry in Pennsylvania," Water Resources Bulletin, Vol. 19, no. 6, 1983, pp. 993-1001.

Excellent report on measurement of acidity in winter snow build-up in the Appalachian Mountains of Pennsylvania. This report has implications for local and regional stream and lake acidity because of the rapid input of snowmelt in spring. Useful data for limnologists and hydroecologists.

213. Dewees, D.N. Evaluation of Policies for Regulating Environmental Pollution, Working Paper No. 4. Ottawa: Economic Council of Canada, 1980.

Good overview and analysis of Canadian environmental policy and regulation procedures. Special emphasis given to Canadian acid rain problem.

214. Dick, W.A. "Organic Carbon, Nitrogen, and Phosphorous Concentrations and pH in Soil Profiles as Affected by Tillage Intensity," Soil Science Society of America Journal, Vol. 47, 1983, pp. 102-107.

A basic knowledge compilation which looks at soil chemistry, especially vertical movement of nutrients, in a variety of agricultural utilization situations.

215. Dickson, W. "Water Acidification--Effects and Countermeasures," Proceedings of the Conference on Ecological Effects of Acidic Deposition. Stockholm: SNSF Project, 1982, pp. 127-144.

One of the most definitive articles on what to do to reverse the effects of acid rain in aquatic ecosystems. Mostly speculative but basic to further discussion.

216. Dillon, P.J., et al. "Acidic Precipitation in South Central Ontario: Recent Observations," Journal of the Fisheries Research Board of Canada, Vol. 35, 1978, pp. 809-815.

Good regional report of field collected data. For comparative purposes.

217. ________. "The Use of Calibrated Lakes and Watersheds for Estimating Atmospheric Deposition Near a Large Point Source," Water, Air, and Soil Pollution, Vol. 18, 1982, pp. 241-250.

A methodological analysis focusing on the factors which can affect lacustrine acidity. Specifically aimed at evaluating lakes as valid indicators of aerosol effluent. Discusses many variables, including local land use patterns.

218. Dixit, S.P. "Influences of pH on Electrophoretic Mobility of Some Soil Colloids," Soil Science, Vol. 133, No. 3, 1982, pp. 144-149.

Background information necessary to an understanding of soil chemistry under the effects of acid rain. Does not deal directly with acid rain but is concerned with soil pH.

219. Dochinger, L.S., et al. "Responses of Hybrid Poplar Trees to Sulfur Dioxide Fumigation," Journal of the Air Pollution Control Association, Vol. 22, 1972, pp. 369-371.

A short report on research with sulfur dioxide fumigation of trees. While this research is "pre acid rain" it provides information on how plants react to one of the constituents of acid rain. Good physiological data provided.

220. Dochinger, L.S. and T.A. Seliga. "Acidic Precipitation and the Forest Ecosystem," Journal of the Air Pollution Control Association, Vol. 22, 1975, p. 11.

A very short report on expected and inferred damage to forest environments from acid rain. No new data presented but a general warning and call for research.

221. Dochinger, L.S. and T.A. Seliga (eds). Proceedings of the First International Symposium on Acid Precipitation and the Forest Ecosystem, Technical Report NE-23. Columbus, OH: USDA Forest Service, 1976.

An early compilation of research reports on acid rain effects in North America. A benchmark for U.S. study.

222. Dohrenwend, R.E., et al. "Acid Precipitation in the Keweenaw Peninsula of Michigan's Upper Peninsula," in D. Drablos, and A. Tollan (eds), Ecological Impact of Acid Precipitation. Oslo: SNSF Project, 1980, pp. 106-107.

A short technical note documenting localized acid rain effects in Michigan. Part of a compendium of scientific information; for technical readers.

223. Dolske, D.A. and F.D. Gatz. "A Field Intercomparison of Sulfate Dry Deposition Monitoring and Measurements Methods: Preliminary Results," in Proceedings of the Acid Rain Symposium. Washington, DC: American Chemical Society, 1982, paper no. 1-33.

Analysis of methods of collecting dry deposition samples: compares accepted methods. No new means suggested. For scientists only.

224. Dolske, D.A. and G.J. Stensland. "A Comparison of Ambient Airborne Sulfate Concentrations Determined by Several Different Filtration Techniques," in U.S. E.P.A., Proceedings; National Symposium on Recent Advances in Pollutant Monitor-

ing of Ambient Air and Stationary Sources, EPA-600/9/84-001. Washington, DC: U.S. Government Printing Office, 1983, pp. 319-328.

A report on measurement of airborne SO_x at a rural Illinois site. Though data are presented that are useful, the focus is on the different types of filtering materials used to sample for pollutants. Good methodological treatise.

225. Dorfman, R. and N.S. Dorfman (eds.). Economics of the Environment--Selected Readings. New York: W.W. Norton and Co., 1977.

Contains some early economic and policy analysis writings on acid rain effects and control in Canada. Only portions are applicable to acid rain.

226. Dovland, H. and A. Eliassen. "Dry Deposition on a Snow Surface," Atmospheric Environment, Vol. 10, 1976, pp. 783-785.

The first study of deposition on snow rather than by or in snow. Dry deposition is a minor phenomenon but this is a good research report on rates of dry deposition.

227. Drablos, D. and A. Tollan (eds). Proceedings of the International Conference on the Ecological Impact of Acid Precipitation. Sandefjord, Norway: SNSF Project, 1980.

A comprehensive collection of papers on many aspects of the acid precipitation problem. Includes sections on atmospheric dynamics and terrestrial effects. International nature makes this work especially valuable.

228. Draxler, R.R. and A.D. Taylor. "Horizontal Dispersion Parameters for Long-range Transport Modeling," Journal of Applied Meteorology, Vol. 21, 1982, pp. 367-372.

A scientific analysis of the portions of upper air circulation that support the movement of acid-forming chemicals over long distances. An examination of some mathematical modeling variables.

229. Driscoll, C.T., et al. "Effect of Aluminum Speciation on Fish in Dilute Acidified Waters," Nature, Vol. 284, 1980, pp. 161-163.

Good technical report on research of metal uptake by fish in waters affected by acid rain or by more direct acidic pollution.

230. DuBay, D.T. and W.H. Murdy. "Direct Adverse Effects of Sulfate on Seed Set in Geranium carolinianum (L): A

Consequence of Reduced Pollen Germination on the Stigma," Botanical Gazette, Vol. 144, no. 3, 1983, pp. 376-381.

A thorough research report on one physiological effect of an acid-rain-related chemical on a typical plant. This begins to fill in the gaps in botanical research into the exact impact acid rain can have on plant communities.

231. ______. "The Impact of Sulfate on Plant Sexual Reproduction: In Vivo and In Vitro Effects Compared," Journal of Environmental Quality, Vol. 12, 1984, pp. 23-29.

Report on continuation of research mentioned above. Comparative nature of these experiments adds complexity and utility to the results.

232. Duncan, J.R. and D.J. Spedding. "Effect of Relative Humidity Adsorption of Sulfur Dioxide Onto Metal Surfaces," Corrosion Science, Vol. 13, 1973, pp. 993-1001.

A highly technical report on a small portion of the chemical reaction involved with metal corrosion by acidic deposition. Very specialized.

233. Durham, J.L. (ed). Chemistry of Particles, Fogs and Rain. Boston: Butterworth Publishers, 1982.

A very technical collection of articles on chemistry and physics which has a bearing on the evolution of atmospheric conditions that lead to acid rain. Not for non-scientific readers.

234. Durran, D., et al. "A Study of Long Range Air Pollution Problems Related to Coal Development in the Northern Great Plains," Atmospheric Environment, Vol. 13, 1979, pp. 1021-1037.

A good analysis of the known and suspected effects of large scale high-sulfur power plant development in the western U.S. Particularly important considering the plans for more Four Corners projects in that region.

235. Durst, C.S., et al. "Horizontal Diffusion in the Atmosphere as Determined by Geostrophic Trajectories," Journal of Fluid Mechanics, Vol. 6, 1959, pp. 401-422.

A scientific examination of the way in which circulation around pressure cells interrelates to move air over long distances. Basic research, not specifically on acid rain.

236. Duvigneaud, P., ed. Productivity of Forest Ecosystems, Paris: Unesco Publications, 1969.

A world view of resource availability and management of forest resources. Does introduce early reference to the effects of pollution on forest ecology.

237. Dvorak, A.J. and B.G. Lewis. Impacts of Coal-fired Power Plants on Fish, Wildlife, and Their Habitats, FWS/OBS 78/29. Washington, DC: U.S. Department of the Interior, 1978.

A general document of federally accepted reports on coal-fired generating plants and local and regional ecology. Deals with acid rain only as a secondary problem.

238. Easter, R.C. and J.M. Hales. Mechanistic Evaluation of Precipitation Scavenging Data Using a One-dimensional Reactive Storm Model, APRI RP-2022-1. Chicago: Battelle Laboratories, 1983.

Very technical but thorough evaluation of one method utilized to predict the effects of combustion chemicals on the atmosphere. Basic to an understanding of the physics and chemistry of acid rain.

239. Eaton, J.S., et al. "Throughfall and Stemflow Chemistry in a Northern Hardwood Forest," Journal of Ecology, Vol. 61, no. 2, 1973, pp. 495-508.

Good technical report on the way in which precipitation reaches the forest floor. Looks at basic mechanisms of transport of precipitation through the canopy. No direct information on acid rain but good background information.

240. ________. "The Input of Gaseous and Particulate Sulfur to a Forest Ecosystem," Tellus, Vol. 30, 1978, pp. 546-551.

Examines the many ways in which chemical pollution is introduced into a forest environment; includes a discussion of acid rain in addition to more direct phenomena.

241. Edgerton, E.S. and P.L. Brezonik. Acid Rainfall in Florida: Results of a Monitoring Network for 1980, Department of Environmental Engineering Sciences monograph. Tallahassee, FL: University of Florida, 1981.

A complete account of a statewide precipitation-monitoring program established in Florida. Data sets are extensive and quite useful for comparison with other regions and for integration into the NADP data.

242. Edinger, J.G. and T.F. Press. Meteorological Factors in the Formation of Regional Haze, Final report for EPA ORD ESRL. Washington, DC: U.S. Environmental Protection Agency, 1982.

A good look at low-level chemical reactions and air mixing. Particularly useful for understanding local pollution problems.

243. Edison Electric Institute. Electricity: Sources and Technologies. Washington, DC: Edison Electric Institute, 1982.

Written for an informed public to present an industrial perspective on some major energy-related questions, including acid rain and other environmental effects of coal combustion. Stresses the idea that "the controlling factors affecting the process" of acid rain production "are not understood very well."

244. Eiler, J.M. and R.G. Berg. Sensitivity of Aquatic Organisms to Acidic Environments. Duluth, MN: Environmental Protection Agency, 1981.

A regional report outlining actual and potential effects of acid rain in Minnesota and surrounding states. Good for informed lay reader.

245. Einbender, G. et al. "The Case for Immediate Controls on Acid Rain," Materials and Society, Vol. 6, no, 3, 1982, pp. 251-282.

An environmentalist's appeal for introduction of short-term policy that can slow deterioration of the acid rain situation while long-term policy is worked out.

246. Electric Power Research Institute. "Tracking the Clues to Acid Rain," EPRI Journal, Sept., 1979, pp. 20-24.

A look at alternate sources for causes of acid rain. Basically a public relations article designed to refocus concern away from electric utilities.

247. Electric Power Research Institute. Ecological Effects of Acid Precipitation, EPRI Report no. EPRI-EA-79-6-LD. Palo Alto, CA: EPRI, 1979.

A comprehensive report of the known effects of acid rain on the environment; the effects are downplayed and any effects which are not absolutely confirmed are left out or blamed on other problems.

248. ________. Acid Deposition: Decision Framework, EPRI Research Report no. EPRI-EA-2540. Palo Alto, CA: EPRI, 1982.

A pro-electric utility analysis of regulatory policy on acid rain control.

249. ______. UAPSP Precipitation Data Displays for January 1, 1979-June 30, 1982, Volume I. Washington, DC: Utility Acid Precipitation Study Program, August, 1983.

An excellent source of raw data on precipitation chemistry. Data are compiled by EPRI as part of their overall research thrust in acid rain.

250. ______. UAPSP Precipitation Data Displays for January 1, 1979-June 30, 1982, Volume II. Washington, DC: Utility Acid Precipitation Study Program, August, 1983.

Identical report to that above.

251. Elgmork, K., et al. "Polluted Snow in Southern Norway During the Winters, 1968-1971," Environmental Pollution, Vol. 4, 1973, pp. 41-52.

A presentation of very useful baseline data which has been and can be an example for inference or for comparative analysis.

252. Elias, R.W. and C.C. Patterson. "The Toxicological Implications of Biogeochemical Studies of Atmospheric Lead," in T.Y. Toribara, et al. (eds), Polluted Rain. New York: Plenum Press, 1980, pp. 391-403.

Proposes research methods for animal tests of effects of anthropogenic aerosol pollutants. This study specific to lead but this and other heavy metals are related in many ways to acid rain production, transport, and effects. Good experimental design proposed for toxicologists.

253. Eliassen, A. and J. Saltbones. "Decay and Transformation Rates of SO_2 as Established From Emission Data, Trajectories, and Measured Air Concentrations," Atmospheric Environment, Vol. 9, 1975, pp. 425-429.

Report of basic research into the chemistry of combustion effluent. Information of the time needed and amounts of acid formed by air pollution is very useful and applicable today.

254. Eliassen, A. "A Review of Long-range Transport Modeling," Journal of Applied Meteorology, Vol. 19, 1980, pp. 231-240.

A compilation, with minimal analysis, of the research on atmospheric transport of effluent available at the time of publication. Dated material but useful for temporal comparison.

255. Elliot, R.A., et al. "SO_2 Emission Control in Smelters," Environmental Progress, Vol. 1, no. 4, 1982, pp. 261-267.

A non-technical report on the ways in which the steel industry has controlled air pollution. This deals with the kind of situation that creates the acid rain problem.

256. Emanuel, W.R., et al. "Modeling Terrestrial Ecosystems in the Global Carbon Cycle with Shifts in Carbon Storage Capacity by Land-use Change," Ecology, Vol. 65, no. 3, 1984, pp. 970-983.

A very insightful look at the way in which one anthropogenic effect, land use patterns, can affect a seemingly unrelated anthropogenic phenomenon, acid precipitation. This is important to developing an approach to softening the effects of airborne pollutants on various ecosystems.

257. Ember, L.R. "Acid Rain Focus of International Cooperation," Chemical and Engineering News, Vol. 57, no. 49, 1979, pp. 15-17.

A general non-technical report of international scientific and political actions with regard to acid rain effects and causes.

258. _______. "States Anguish Over Acid Rain Problems," Chemical and Engineering News, Vol. 58, no. 16, 1979, pp. 22-23.

A journalistic report of local and regional concern for acid rain in their area.

259. _______. "Power Plants Cited as Big Acid Rain Source," Chemical and Engineering News, Vol. 59, no. 38, 1981, pp. 14-15.

Non-technical, journalistic report of political accusations against electric power plants.

260. _______. "Acid Rain Implicated in Forest Dieback," Chemical Engineering News, Vol. 60, no. 47, 1982, pp. 25-26.

Another short, journalistic report chronicling research on forest effects. Good for lay audience.

261. Environmental Science and Engineering, Inc. Phase II Report Florida Acid Deposition Study (Plan of Study), ESE no. 83-152-600. Gainesville, FL: ESE, 1983.

A consultant report on a plan to establish ongoing monitoring and analysis of acid deposition in Florida. Primarily research design and policy decisions, this document will be extremely useful as a model for establishing similar networks in other states or regions.

262. Eriksson, E. "Composition of Atmospheric Precipitation," Tellus, Vol. 4, 1952, pp. 215-232.

An in-depth overview of aerosol effluents, including acidic compounds. Perhaps too much other information if interested only in acid rain.

263. Eriksson, E. "The Yearly Circulation of Sulfur in Nature," Journal of Geophysical Research, Vol. 68, 1963, pp. 4001-4008.

A "pre-acid rain" examination of the natural movement of sulfur through the upper air circulation system. Helpful to an understanding of the basic chemical and physical processes that affect acidification of the atmosphere.

264. Eriksson, M.O.G., et al. "Predator-Prey Relations Important for the Biotic Changes in Acidified Lakes," Ambio, Vol. 9, 1980, pp. 248-249.

Short research report on the realignment of food chains in acidified lakes. Very important article for lacustrine ecology and studies of ecosystem regeneration.

265. Eschenroeder, A. "Atmospheric Dynamics of NO_x Emission Controls," in M.M. Benarie (ed), Atmospheric Pollution 1982. New York: Elsevier Publishing Company, 1982, pp. 71-90.

A report on the fundamental physics and chemical processes influencing the control of nitrogen oxide emission from combustion sites. Criteria for control to meet air quality standards are discussed and analyzed with respect to these processes.

266. Estoque, M.A., et al. "A Lake Breeze Over Southern Lake Ontario," Monthly Weather Review, Vol. 104, 1976, pp. 386-396.

Basic scientific report on diurnal and seasonal regional low-level circulation in an area of intense concern with regard to acid rain. Very helpful in understanding the regional circulation patterns that underlie international transportation of chemicals in North America.

267. Eubanks, L.S. and R.A. Cabe. "Normative Economics and the Acid Rain Problem," in T.D. Crocker (ed), Economic Perspectives on Acid Deposition Control. Boston: Butterworth Publishers, 1984, pp. 81-96.

A very technical econometric analysis of the pros and cons to effluent control and acid rain effects.

268. Evans, L.S. et al. "Leaf Surface and Histological Perturbations of Leaves of Phaseolus Vulgaris and Helianthus annus After Exposure to Simulated Acid Rain," American Journal of Botany, Vol. 64, 1977, p. 903.

An early report on the effects of acidified precipitation on important agricultural plants. Results are basic to an understanding of the ways in which acid rain will or will not affect agriculture.

269. _______. "Foliar Response of Six Clones of Hybrid Poplar to Simulated Acid Rain," Phytopathology, Vol. 68, 1978, pp. 847-856.

A very technical report of research on leaf damage caused by simulated acid rain on a common forest tree. Results are mixed but seem to indicate a minimal amount of damage to the overall health of the plant. Good source of comparative data and good analysis.

270. Evans, L.S. "Foliar Responses That May Determine Plant Injury," in T.Y. Toribara, et al. (eds), Polluted Rain. New York: Plenum Press, 1980, pp. 239-257.

A most thorough report of experiments with simulated acid rain on various plants (including ferns and two types of beans) to isolate foliar responses that may indicate further damage to plants. Overall results indicate that foliar damage is localized although it may be quite intense. Good source of physiological data and analysis.

271. _______ and K.F. Lewin. "Effects of Simulated Acid Rain on Growth and Yield of Soybeans and Pinto Beans," in D.S. Shriner, et al., (eds), Atmospheric Sulfur Deposition. Ann Arbor, MI: Ann Arbor Science Publishers, 1980, pp. 299-307.

Documents acid rain effects on foliage as well as indirect effects through soil pollution. Includes primary research by the authors as well as reports from other scientists. Very good information for informed lay readers as well as scientists.

272. _______, et al. "Acidic Precipitation: Considerations for an Air Quality Standard," Water, Air, and Soil Pollution, Vol. 16, 1981, pp. 469-509.

An agricultural scientist speaks out on proposed policy. Good insights combined with some interesting naiveté.

273. _______. "Biological Effects of Acidity in Precipitation on Vegetation: A Review," Environmental and Experimental Botany, Vol. 22, no. 2, 1982, pp. 155-169.

A good modern synopsis of research into this subject. No other compendium as comprehensive or readable.

274. Everett, A.G., et al. Hydro Geochemical Characteristics of Adirondack Waters Influenced by Terrestrial Environments, consultant report by Everett and Associates, Rockville, Maryland, December 1982.

A technical report of the ways in which local geology, land use patterns, aerosol pollution, and local water supplies interact in a very sensitive region. The geological vulnerability of the Adirondack region to acid precipitation is clearly illuminated.

275. Fairfax, J.A.W. and N.W. Lepp. "Effect of Simulated 'Acid Rain' on Cation Loss From Leaves," Nature, Vol. 255, 1975, pp. 324-325.

Short technical report on ionic activity on and near leaf surfaces under the influence of acidic precipitation. Particularly useful in assessing what, if any, direct foliar effects acid deposition inflicts. At a physiological level.

276. Falconer, R.E. and P.D. Falconer. Determination of Cloud Water Acidity at a Mountain Observatory in the Adirondack Mountains of New York, Publication 741. Albany, NY: State University of New York, 1979.

A comprehensive compilation of data on measuring atmospheric pH without traditional sampling systems. Complete listing of data provided.

277. ________. "Determination of Cloud Water Acidity at a Mountain Observatory in the Adirondack Mountains of New York State," Journal of Geophysical Research, Vol. 85, 1980, pp. 7465-7470.

A rare analysis of acid rain in pre-precipitation state. Very useful basic meteorological report in a critical U.S. region. Good source for atmospheric chemists and physicists.

278. Fay, J.A. and J.J. Rosenweig. "An Analytical Diffusion Model for Long-distance Transport of Air Pollutants," Atmospheric Environment, Vol. 14, 1980, pp. 355-365.

Technical report of development of a mathematical technique for predicting dispersal of effluent in a tall stack situation. A refinement of several other models of plume dispersion so a good source of comparative information and analysis.

279. Fay, J.A. Long-Range Transport of Acid Rain Precursors, Report No. MIT-EL 83-005. Cambridge, MA: MIT Energy Laboratory, 1983.

A solid technical report of an analysis of the movement of combustion effluent through the atmosphere across long distances. Good basis for modeling.

280. Feagley, S.E. and R.B. Crenmers. "Acid Rain and Its Accumulation: A Problem in Louisiana?" Louisiana Agriculture, Vol. 24, 1984, pp. 12-14.

A rather general article meant to introduce regional effects of acid rain to the people of Louisiana. Highlights some of the research ongoing in the School of Agriculture at Louisiana State University but only briefly.

281. Felske, B.E. Sulphur Dioxide Regulation and the Nonferrous Metals Industry, Technical Report No. 3. Ottawa: Economic Council of Canada, 1981.

Primarily an economic analysis of the effects of current and proposed regulation on large smelting activities in Canada. Very useful for policy and regulatory analysts as a comparative model of action.

282. Ferenbaugh, R.W. "Effects of Simulated Acid Rain on Phaseolus Vulgaris L.," American Journal of Botany, Vol. 63, 1976, pp. 283-288.

A specialized report on laboratory experiments with an agricultural plant under the effects of simulated acidic precipitation. Early report but results are still useful.

283. Fisher, D.W., et al. "Atmospheric Contributions to Water Quality of Streams in the Hubbard Brook Experimental Forest, New Hampshire," Water Resources Research, Vol. 4, no. 5, 1968, pp. 1115-1126.

A concise report of one aspect of the longitudinal field and laboratory study still active in Hubbard Brook. Very good early data on water resources under attack from acid rain.

284. Flack, W.W. and M.J. Matteson. "Mass Transfer of Gases to Growing Water Droplets," in T.Y. Toribara, et al. (eds), Polluted Rain. New York: Plenum Press, 1980, pp. 61-85.

An analysis of the way in which pollutants are introduced into the precipitation formation process in the atmosphere. Separate analyses and descriptions for SO_2, NO_2 and O_2. No new information but a good overview of the chemistry of acid rain formation.

285. Florida Department of Environmental Regulation. An Analysis of Acid Deposition Issues: The Impacts of Proposed National Acid Deposition Control Legislation on Florida. Tallahasee, FL: Florida Public Service Commission, 1984.

A rather bureaucratic report that does provide some insight into local reaction and concern in response to national legislation of environmental issues. Particularly useful because Florida is a state that depends upon coal-generated power and one that contains many sensitive environmental systems.

286. Folkeson, L. The Use of Ecological Variables in Environmental Monitoring, PM 1151,59. Stockholm: National Swedish Environmental Protection Board, 1979.

A methodological treatise that can be of use to ecologists in designing experiments or data gathering field studies. Also useful in evaluating other research.

287. Forland, E.J. "A Study of the Acidity in the Precipitation in Southwestern Norway," Tellus, Vol. 25, 1973, pp. 291-298.

One of the first modern studies of acid rain in Europe. This article provides baseline data that became part of the evidence that led to efforts to control acidity in Europe. Data are still useful.

288. Forster, B.A. "A Note on Economic Growth and Environmental Quality," Swedish Journal of Economics, Vol. 56, 1972, pp. 281-285.

A short non-technical evaluation of the way in which expansion of technological societies is tied to environmental pollution. Good for policy analysis and for philosophical discussion.

289. ________. "Economic Impact of Acid Precipitation: A Canadian Perspective," in T.D. Crocker (ed), Economic Perspectives on Acid Deposition. Boston: Butterworth Publishers, 1984, pp. 97-122.

An excellent analysis of the ways in which both the production of pre-acidic effluent in the manufacturing sector and the effects of acidic precipitation are socio-economic problems as well as environmental and ecological ones. Good data for policy development.

290. Fowler, D. "Removal of Sulfur and Nitrogen Compounds from the Atmosphere in Rain and by Dry Deposition," in D. Drablos and A. Tollan (eds), Ecological Impact of Acid Precipitation. Sandefjord, Norway: SNSF Project, 1980, pp. 22-32.

A comprehensive overview of pollutant washout and dropout focusing on the two most common chemicals in acid rain. Provides some good data sets as well as a thorough analysis of the processes involved with removal of these chemicals from the atmosphere.

291. Francis, A.J., et al. "Effect of Lake pH on Microbial Decomposition of Allochthonous Litter," in G.R. Hendrey (ed), Early Biotic Responses to Advancing Lake Acidification. Boston: Butterworth Publishers, 1984, pp. 1-21.

A very scientific examination of the cycling of carbon through the microorganismic ecosystems in freshwater lakes and the effects of acidity on both the microorganisms and on the carbon cycle. By focusing on the effects of acidity on organic decomposition of leaves deposited in lakes, the authors illuminate the influence acid rain can have on one of the basic segments of lacustrine ecology. Increased acidity results in decreased microorganic activity in this situation.

292. Freeman, A.M. "Air and Water Pollution Policy," in P.R. Portney (ed), Current Issues in U.S. Environmental Policy. Baltimore, MD: Johns Hopkins University Press, 1978, pp. 12-67.

A discussion of the logic and philosophy behind development of environmental policy in the U.S. This looks at federal vs state responses, responsibilities, the need for enforcement at both levels, and the way in which penalties for non-compliance are invoked. Gives considerable attention to Clean Air Act programs including acid rain.

293. Friedman, R.M., et al. "The Acid Rain Controversy: The Limits of Confidence," American Statistician, Vol. 37, no. 4, 1983, pp. 385-394.

A technical analysis of the data gathering and data manipulation and interpretation problems that have been created by the various precipitation and effluent monitoring programs in the U.S.

294. Frohliger, J.O. and R. Kane. "Precipitation: Its Acidic Nature," Science, Vol. 189, 1975, pp. 455-457.

A short discussion of natural and anthropogenic acidity in precipitation. A general article, not a result of new research.

295. Fromholzer, D. Decision Framework for Ambient Air Quality Standards: Methodology and Model Structure, EA-3193. Palo Alto, CA: Electric Power Research Institute, 1984.

An industry-oriented evaluation of the policy and regulation development process in the United States. Evaluates the current decision-making process and makes substantive recommendations for changes in that process. Very useful for policy students.

296. Fromholzer, D. and R. Richels. Acid Deposition Decision Framework: State-level Applications. Palo Alto, CA: Electric Power Research Institute, 1984.

A monograph which examines the various ways in which state governments can and do react to local acid deposition problems. A very thorough analysis of state-level public policy; the ways in which states view economic reaction to acid rain regulation is particularly interesting. Of much interest to students of public policy.

297. Fromm, P.O. "A Review of Some Physiological and Toxicological Responses of Freshwater Fish to Acid Stress," Environmental Biology, Vol. 5, no. 1, 1980, pp. 79-93.

A general overview of research results in icthic ecology. Rather scientific at times but useful and understandable for informed lay audience.

298. Fuquay, J.J. "Scavenging in Perspective," in R.J. Engelmann and W.G.N. Slinn (eds), Precipitation Scavenging. Washington, DC: Atomic Energy Commission, 1970, pp. 22-31.

An overview of one atmospheric process that is a contributor to acid rain formation. Not focused on acid rain per se but basic to an understanding of the development of the modern problem.

299. Galloway, J.N., et al. "Acid Precipitation in the Northeastern United States," Science, Vol. 194, 1976, p. 217.

One of a series of general presentations aimed at a non-specific scientific audience. No new information but useful graphics and interpretations.

300. Galloway, J.N. and E.V. Cowling. "The Effects of Precipitation on Aquatic and Terrestrial Ecosystems: A Proposed Precipitation Chemistry Network," Journal of Air Pollution Control Association, Vol. 28, 1978, pp. 228-235.

An evaluation of the precipitation data problem at an early stage in U.S. acid rain concern. This article helped lead to creation of a national system of data collection. Good for policy analysts.

301. Galloway, J.N., et al. A National Program for Assessing the Problem of Atmospheric Deposition (acid rain). Fort Collins, CO: Natural Resource Ecology Laboratory, 1978.

Statement on the organizational structure and data collection strategy that lies behind the largest U.S. precipitation chemistry monitoring program. Good for policy study.

302. Galloway, J.N. and G.E. Likens. "The Collection of Precipitation of Chemical Analysis," <u>Tellus</u>, Vol. 30, 1978, pp. 71-82.

A good overview of methodological problems associated with monitoring precipitation chemistry both on a specific project basis and on a national or international level.

303. Galloway, J.N., et al. "Acid Precipitation: Measurement of pH and Acidity," <u>Liminological Oceanography</u>, Vol. 24, no. 6, 1979, pp. 1,161-1,165.

A general overview of the problem of acid rain from a water resource perspective. Some discussion of data collection methodology.

304. Galloway, J.N. and G.E. Likens. "Atmospheric Enhancement of Metal Deposition in Adirondack Lake Sediments," <u>Limnological Oceanography</u>, Vol. 24, 1979, pp. 427-433.

Technical report on the way in which acid rain can affect heavy metal activity in lacustrine environments.

305. Galloway, J.N. and D.M. Whelpdale. "An Atmospheric Sulfur Budget for Eastern North America," <u>Atmospheric Environment</u>, Vol. 14, 1980, pp. 409-417.

Rather technical presentation of atmospheric chemistry of a major chemical component of acid rain. General but scientifically useful.

306. Galloway, J.N., et al. "Changing pH and Metal Levels in Streams and Lakes in the Eastern United States Caused by Acid Precipitation," in <u>Restoration of Lakes and Inland Waters</u>, EPA Report 440/81-010. Washington, DC: U.S. Environmental Protection Agency, 1980, pp. 446-453.

An excellent scientific report of research into one set of indirect effects created by acid rain. Technically sound but usable by nonscientists also.

307. Galloway, J.N. and G.E. Likens. "Acid Precipitation: The Importance of Nitric Acid," <u>Atmospheric Environment</u>, Vol. 15, 1981, pp. 1081-1085.

Single substance analysis of the unrelated and related chemical effects of one component of acid rain. Good basic scientific analysis.

308. Galloway, J.N., et al. <u>Shenandoah Watershed Acidification Study: Three Year Summary Report, September 1979-September 1982</u>. Denver, CO: National Park Service, 1982.

A thorough and wide-ranging report on the effects of acid precipitation in a pristine natural environment. Includes data and analysis on precipitation trends, water resources chemistry, fish and wildlife reactions to acidification, and soil and forest vegetation reaction trends. A model for scientific evaluation on a small regional scale.

309. Galloway, J.N., et al. "Freshwater Acidification from Atmospheric Deposition of H_2SO_4--A Conceptual Model," Environmental Science and Technology, Vol. 17, no. 11, 1983, pp. 334-343.

A very technical presentation of a predicted scenario of hydrologic acidification with one major component of acid rain. Excellent documentation of the processes involved and an in-depth analysis. For scientists only.

310. Galloway, J.N., et al. "Acid Precipitation: Natural vs Anthropogenic Components," Science, Vol. 226, no. 4676, 1984, pp. 829-831.

A rather short article addressing the problem of identifying anthropogenic acidity. The authors document the regional increases in H, SO_4, and NO_3 in precipitation since 1950 as evidence. They also discuss the dramatic increases in acidity in industrial rather than non-industrial regions of the U.S. Good source of general data for a non-scientific audience.

311. Galvin, P.T., et al. "Transport of Sulfate to New York State," Environmental Science and Technology, Vol. 12, no. 5, 1978, pp. 580-584.

An analysis of the atmospheric transportation process which deposits sulfur compounds through precipitation in New York state. More descriptive case study than atmospheric model. Very useful for developing regional models of acidification.

312. Gambell, A.W. and D.W. Fisher. Chemical Composition of Rainfall in Eastern North Carolina and Southeastern Virginia, U.S. Geological Survey Water Supply Paper 1535-K. Washington, DC: U.S.G.S., 1966.

Compilation of data collected for the Eastern U.S. several years before the acid rain problem became a national focus of interest. Usable data and graphics. Especially useful for temporal comparison with modern data.

313. Garland, J.A. "Field Measurements of the Dry Deposition of Small Particles to Grass," in H.W. Georgii and J. Pankrath (eds), Deposition of Atmospheric Pollutants. Boston: Reidel Publishing, 1982, pp. 9-16.

A report on a field test of dry deposition rates previously measured in a wind tunnel. Rates of deposition of lead, Aitken nuclei and small monodisperse particles on grass are very similar to wind tunnel results. Good set of data for comparative study.

314. Gartrell, F.E., et al. "Atmospheric Oxidation of SO_2 in Coal-burning Power Plant Plumes," American Industrial Hygiene Association Journal, Vol. 24, 1963, p. 113.

A short research note on the rate at which SO_2 is turned into acidic compounds in the atmosphere near a combustion site. Good background information for an overall understanding of the acid rain process.

315. Gasper, D.C. The West Virginia "Acid Rain" Fishery Problem. Charleston, WV: Department of Natural Resources, 1983.

An examination of the effects of acid rain on fish resources in West Virginia. The problem of acid rain effects is clouded by the problem of coal mine drainage across the state but an attempt is made to delineate the two sets of effects.

316. Gatz, D.F. A Review of Chemical Tracer Experiments on Precipitation Systems," Atmospheric Environment, Vol. 11, 1977, pp. 945-953.

A good overview of research into the atmospheric movement of chemicals that constitute acid rain.

317. ________. "Identification of Aerosol Sources in the St. Louis Area Using Factor Analysis," Journal of Applied Meteorology, Vol. 17, 1978, pp. 600-608.

Very technical presentation of a model for detecting pollution sources from non-source data. Part of the comprehensive study done in the St. Louis region to develop a regional air pollution emission and transport model.

318. Gauri, K.L. "Deterioration of Architectural Structures and Monuments," in T.Y. Toribara, et al. (eds), Polluted Rain. New York: Plenum Press, 1980, pp. 125-145.

A thorough examination of the ways in which acid rain can affect concrete, stone, and mortar structures. Provides information on breakdown rates for several major minerals contained in popular building materials under acid rain attack.

319. Georgii, H.W. "Global Distribution of the Acidity in Precipitation," in H.W. Georgii and J. Pankrath (eds), Deposition of Atmospheric Pollutants. Boston: Reidel Publishing, 1982, pp. 55-66.

An attempt to use World Meteorological Organization data from 1972-79 to develop a spatial model of precipitation pH. The data really only support such work in Europe and in North America. Good for methodological discussion and a good source of data and contour maps of the two regions.

320. _______ and J. Pankrath (eds). Deposition of Atmospheric Pollutants. Boston: Reidel Publishing, 1982.

A good collection of articles taken from a colloquium held at Oberursel/Taunus, GDR in 1981. Provides a very international look at various aspects of acid rain and related topics. Main headings are Wet Deposition, Dry Deposition, and Effects on Plants.

321. Georgii, H.W., et al. "Wet and Dry Deposition of Acidic and Heavy Metal Components in the Federal Republic of Germany," in S. Beilke and A.J. Elshout (eds), Acid Deposition. Boston: Reidel Publishing, 1983, pp. 142-148.

Report on precipitation sampling in Germany at thirteen sites in 1979-81. Provides good data sets on major pre-acid chemicals as well as on heavy-metal aerosols. Very little analysis but usable graphical presentations of data.

322. Gerhardt, P.H. "Improving Air Quality--Some Strategy Alternatives," in J.R. Conner and E. Loehman (eds), Economics and Decision Making for Environmental Quality. Gainesville, FL: University Presses of Florida, 1974, pp. 77-95.

This provides a legalistic examination of the economics of pollution both from a societal cost standpoint and from an industrial perspective, focusing on the costs of clean-up. Good general reading; not econometric so it is understandable by most. Policy background.

323. Gholz, H.S. "Effects of Atmospheric Deposition on Forested Ecosystems in Florida--Suggested Research Priorities," in A.E.S. Green and W.H. Smith (eds), Acid Deposition Causes and Effects: A State Assessment Model. Gainesville, FL: University of Florida, 1983, pp. 34-39.

Provides some general statements about the effects of acid precipitation in Florida but concentrates its efforts in proposing areas where data are in short supply. Provides good insight into acid rain research priorities.

324. Gibson, J.H. and R.A. Linthurst. "Effects of Acidic Precipitation on the North American Continent," in T. Schneider and L. Grant (eds), Air Pollution by Nitrogen Oxides. Amsterdam: Elsevier Scientific Publishing Co., 1982, pp. 577-594.

A well-presented overview of the whole spectrum of effects from acid rain. No special focus or in-depth analysis of any particular effect but a very comprehensive treatment.

325. Gjessing, E.T., et al. "Effects of Acid Precipitation on Freshwater Chemistry," in F.H. Braekke (ed), Summary Report on the Research Results From the Phase I (1972-1975) of the SNSF Project, Agricultural Research Council of Norway, 1976, pp. 65-85.

A landmark report of international research that became the basis of European and U.N. policies regarding the control of acid rain.

326. Glass, N.R., et al. "The Ecological Effects of Atmospheric Deposition," EPA Energy/Environment III, EPA-600/9-78-002. Washington, DC: U.S. Environmental Protection Agency, 1978, pp. 113-117.

Short, general statement about acid rain and the environment.

327. Glass, N.R. "Effects of Acid Precipitation," Environmental Science and Technology, Vol. 36, 1979, p. 1352.

A general article intended for a scientific audience. No new research, synthetic in nature but relatively comprehensive.

328. ________, et al. "Effects of Acid Precipitation in North America," Environment International, Vol. 4, 1980, pp. 443-452.

An overview article intended for a general informed public audience. No special emphasis on any particular area of concern; no new research reported.

329. ________, et al. "The Sensitivity of the United States Environment to Acid Precipitation," in Proceedings of the International Conference on the Ecological Impacts of Acid Precipitation. Sandefjord, Norway: SNSF Project, 1980, pp. 119-122.

A general examination of the regions and specific types of ecosystems in the United States that are susceptible to acid rain effects. No new data but an interesting national assessment.

330. ________, et al. "Effects of Acid Precipitation," Environmental Science and Technology, Vol. 16, no, 3, 1982, pp. 162-169.

A slightly technical overview of acid rain by a scientist who has been a leader in acid rain research from the beginning of interest in the subject in the United States. One of the best general articles available.

331. Glover, G.M., et al. "Ion Relationships in Acid Precipitation and Stream Chemistry," in T.C. Hutchinson and M. Havas (eds), Effects of Acid Precipitation on Terrestrial Ecosystems. New York: Plenum Press, 1980, pp. 95-109.

An analysis of the chemical activity in the formation and deposition of acid precipitation. No new data but new analyses provided. Very useful for a technical reader.

332. Goldstein, R.A. Evaluation of Simulated Acid Precipitation Effects on Forest Ecosystems. Palo Alto, CA: Electric Power Research Institute, 1984.

A thorough, current report on simulation research in forest environments. The focus is primarily on plants but there are also data and analysis of effects on soil and forest microbial communities reported. A good source of data for all ecologists.

333. _______. Integrated Lake-Watershed Acidification Study, EA-3221. Palo Alto, CA: Electric Power Research Institute, 1984.

Report on an EPRI-funded study of lacustrine acidification. Considerable emphasis is placed on the non-acid-rain variables (geology, drainage patterns, natural acidity) which play a part in controlling the acidity of lakes. A good, comprehensive research report. Useful for ecologists, hydrologists, and limnologists.

334. Gorham, E. "Atmospheric Pollution by Hydrochloric Acid," The Quarterly Journal of the Royal Meteorological Society, Vol. 84, 1958, pp. 274-276.

A short general statement on acid rain. No new data even for this early time but important because it was reported well before international concern grew.

335. _______ and W.W. McFee. "Effects of Acid Deposition Upon Outputs from Terrestrial to Aquatic Ecosystems," in T.C. Hutchinson and M. Havas (eds), Effects of Acid Precipitation on Terrestrial Ecosystems. New York: Plenum Press, 1980, pp. 465-480.

One of the few reports on the effects of movement of acid rain through one ecosystem into another. Very useful because of the information about the effects of percolation, groundwater flow, and land use on acid rain and vice versa.

336. Gorham, E. "What to do About Acid Rain," Technology Review, Vol. 85, no. 7, 1982, pp. 58-63.

Inferential article dealing with both policy and technological recommendations for remedies.

337. Graham, R.C., et al. "Automated Acquisition, Filtering and Reduction of Ion Chromatographic Data," in U.S. E.P.A., Proceedings; National Symposium on Recent Advances in Pollutant Monitoring of Ambient Air and Stationary Sources, EPA-600/9-84-001. Washington, DC: U.S. Government Printing Office, 1983, pp. 276-289.

A report on a new procedure to speed chemical analysis of acidified precipitation samples. Provides information on instrumentation, procedures, and data handling. Certification of data is a growing problem in acid rain research; this article provides some good suggestions on methodological innovations.

338. Grahn, O., et al. "Oligotrophication--A Self-accelerating Process in Lakes Subjected to Excessive Supply of Acid Substances," Ambio, Vol. 3, 1974, pp. 93-94.

Good basic research report on the large-scale processes affected by acidic precipitation. Very important because this documents the way in which acidification interferes with a natural process in a lacustrine environment.

339. Grahn, O. "Fishkills in Two Moderately Acid Lakes Due to High Aluminum Concentrations," in D. Drablos and E. Tolan (eds) Proceedings, International Conference on Ecological Impacts of Acid Precipitation. Sandefjord, Norway: SNSF 1980. pp. 310-311.

Short research report of monitored fish kills and analysis of fish chemistry.

340. Granat, L. "On the Relation Between pH and the Chemical Composition in Atmospheric Precipitation," Tellus, Vol. 24, 1972, pp. 550-560.

A look at the way in which changes in atmospheric pH affects other atmospheric chemical cycles and processes. Good because of its technical content and because of the unique focus.

341. ________ and H. Rodhe. "A Study of Fallout by Precipitation Around an Oil-fired Power Plant," Atmospheric Environment, Vol. 7, 1973, pp. 781-792.

One of the first scientific articles on a case study of localized acid rain production from a modern anthropogenic source.

342. Gravenhorst, G. and K.D. Hofken. "Concentration of Aerosol Constituents Above and Beneath a Beech and a Spruce Forest Canopy," in H.W. Georgii and J. Pankrath (eds), Deposition of Atmospheric Pollutants. Boston: Reidel Publishing, 1982, pp. 187-190.

Report of research on the filtering efficiency of beech and spruce forests in Europe. Seasonal efficiency of the deciduous trees is demonstrated as is the effect of pollutant particle size on throughfall. Data on heavy metals as well as acid rain constituents collected and in graphic form.

343. Gravenhorst, G., et al. "The Influence of Clouds and Rain on the Vertical Distribution of Sulfur Dioxide in a One Dimensional Steady-state Model," Atmospheric Environment, Vol. 12, 1978, p. 691.

A short research note on the effect of acid rain on atmospheric chemistry.

344. Graves, C.K. "Rain of Troubles," Science 80, Vol. 1, no. 6, 1980, pp. 74-79.

A general, educated-public oriented article on acid rain. No new information but a good overview with usable graphics.

345. Great Lakes Water Quality Board. Report to the International Joint Commission. Chicago: GLWQB, 1976.

The first report on development of an international policy for management of the environment of the Great Lakes. Especially useful for policy analysis and policy development study.

346. _______. Report to the International Joint Commission, Chicago, 1979.

A general report of the progress in policy development for a joint U.S.-Canada agreement on managing the Great Lakes. Particularly useful for policy analysts.

347. Green, A.E.S. and W.H. Smith (eds). Acid Deposition Causes and Effects: A State Assessment Model. Gainesville, FL: University of Florida, 1983.

An interesting collection of articles with a specific regional focus on one state. Contains reports on scientific research on effects and precipitation process as well as public policy development and implementation.

348. Griffin, C.W., Jr. "America's Airborne Garbage," in P.R. Campbell and J.L. Wade (eds), Society and Environment: The Coming Collision. Boston: Allyn and Bacon, 1972, pp. 145-154.

A general, non-scientific examination of the overall problem of air pollution in the U.S. More policy oriented than process or effects related. Does examine the overall problem of downwind transport of pollutants but only from a general policy level.

349. Gruhl, J. Issues Relating to the Costing of Acid Rain Control Technologies. Tucson, AZ: Gruhl Associates, 1983.

A consultant report outlining the various economic, technological, political, and environmental issues involved with controlling acid rain. Unique because it takes an individual utility approach rather than systematic or regional. Useful to a wide range of interests.

350. Haines, B. "Air Pollution in the Southeast United States: A Brief Review," Georgia Journal of Science, Vol. 37, 1979, pp. 185-191.

A general overview of precipitation chemistry in the southeast U.S. and data on related anthropogenic pollutants. Data sets are rather general but the analysis is insightful.

351. _______ and S. Hendrix. "Acid Rain: Plant Species From a Southern Appalachian Forest Succession," Water, Air, and Soil Pollution, Vol. 14, 1980, pp. 403-407.

An interesting report on plant community reactions to acid precipitation in a major forest region. The evidence for overall forest evolution can have implications for other regions and perhaps for other ecosystems.

352. Haines, B. and J. Waide. "Predicting Potential Impacts of Acid Rain on Elemental Cycling in a Southern Appalachian Deciduous Forest at Coweeta," in T.C. Hutchinson and M. Havas (eds), Effects of Acid Precipitation on Terrestrial Ecosystems. New York: Plenum Press, 1980, pp. 114-122.

A thorough report on the known and predicted effects of acid precipitation on the movement and availability of minerals in forest soils. Also reports on potentially damaging heavy metals that become more conspicuous under acid conditions.

353. Haines, B. "Forest Ecosystem SO_4-S Input-Output Discrepancies and Acid Rain: Are They Related," Oikos, Vol. 41, 1983, pp. 1329-1343.

A report on examination of the ionic and compound dynamics in forest soils with an eye toward isolating natural cycles from those induced or created by acid precipitation. Of considerable interest to soil and forest scientists.

354. Haines, T.A. "Acid Precipitation and Its Consequences for Aquatic Ecosystems: A Review," Transactions of the American Fisheries Society, Vol. 110, no. 6, 1981, pp. 669-670.

A short, general overview of the implications of acid rain for lakes and streams.

355. ________. "Interpretation of Aquatic pH Trends," Transactions of the American Fisheries Society, Vol. 111, no. 6, 1982, pp. 781-786.

Good overview of ecological implications of acid rain studies on lacustrine and riverine environments. Of use to aquatic ecologists and hydrologists.

356. Hakkarinen, C.S. "Past and Ongoing Network in Comparison Studies," in D.H. Pack and A.A. Shepherd, (eds), Proceedings: Advisory Workshop on Methods for Comparing Precipitation Chemistry Data. Washington, DC: Utility Acid Precipitation Study Program, 1982, pp. 4-25--4-44.

An analysis of the statistical validity of acid rain data gathering in the United States. Results are slightly slanted toward the position taken by American electric utility companies.

357. Hales, J.M., et al. Field Investigation of Sulfur Dioxide Washout from the Plume of a Large, Coal-fired Power Plant by Natural Precipitation, EPA 22-69-150. Washington, DC: U.S. Environmental Protection Agency, 1971.

An early, very comprehensive examination of the chemistry of the introduction of combustion effluent, specifically sulfur, into the precipitation cycle. Excellent technical quantitative models are developed as well as extensive analysis of this case study.

358. Hales, J.M. "Fundamentals of the Theory of Gas Scavenging by Rain," Atmospheric Environment, Vol. 6, 1972, pp. 634-636.

A short statement of the basic laws and processes involved with pollutant scavenging in the precipitation process. Good concise statement for background information.

359. ________, et al. "A Linear Model for Predicting the Washout of Pollutant Gases From Industrial Plumes," AICHE Journal, Vol. 19, no. 2, 1973, pp. 292-297.

Good technical article that outlines one method for predicting acid rain production from a given point-source. Some weaknesses due to lack of modern data, but very useful to scientists of all areas. Although the author refers to this as a "simplified" method, the article is very technical and recommended for engineers or physical chemists only. An excellent article.

360. Hales, J.M. and M.T. Dana. "Regional-Scale Deposition of

Sulfur Dioxide by Precipitation Scavenging," Atmospheric Environment, Vol. 13, 1979, pp. 1121-1132.

A report on an important subset of the acid precipitation process in which falling precipitation becomes more acidic as it falls through polluted air. This process has been discussed in other articles but never with a regional model as is here introduced.

361. Hales, J.M. "Mechanistic Analysis of Precipitation Scavenging Using a One-dimensional, Time-variant Model," Atmospheric Environment, Vol. 16, 1982, pp. 1775-1783.

Very technical presentation of a proposed model for analyzing one component of the acid rain cycle.

362. _______. "Precipitation Chemistry: Its Behavior and Its Calculation," in S.V. Krupa and A.H. Legge (eds), Air Pollutants and Their Effects on the Terrestrial Ecosystem. New York: John Wiley, 1983, pp. 145-167.

A good overview of modeling of atmospheric chemistry when examining acid rain. The subject is very technical but the presentation makes the information available to non-technicians as well.

363. Hall, R.J. and G.E. Likens. "Ecological Effects of Whole-Stream Acidification," in D.S. Shriner, et al. (eds), Atmospheric Sulfur Deposition. Ann Arbor, MI: Ann Arbor Science Publishers, 1980, pp. 443-451.

A scientific analysis which develops a model for what will happen to an entire stream system when acidified. Result is based on projections from actual data in specific locations.

364. Hammer, U.T. Acid Rain: The Potential for Saskatchewan. Regina: Saskatchewan Environmental Advisory Council, 1980.

A predictive overview of environmental, economic, and social effects of continued acid precipitation. Specific to this region of Canada but applicable to other parts of the continent. General, non-technical.

365. Hane, C.E. "Scavenging of Urban Pollutants by Thunderstorm Rainfall: Numerical Experimentation," Journal of Applied Meteorology, Vol. 17, 1978, pp. 699-710.

Very technical examination of possible models for urban, non-point source generation of atmospheric pollutants that may lead to acid rain.

366. Hansen, D.A., et al. Examination of the Bases for Trend

Interpretation of Historical Rain Chemistry in the Eastern United States. Washington, DC: Utility Air Regulatory Group Committee on Acid Precipitation, ERT Doc. No. P-A097R, 1981.

A utility-oriented group looks at rain chemistry in an attempt to find trends that do not reflect culpability by utility generation.

367. Hansen, D.A. and G.M. Hidy. "Review of Questions Regarding Rain Acidity Data," Atmospheric Environment, Vol. 16, 1982, pp. 1917-1925.

A retort to questions raised by the electric utility organizations on the validity of acid rain data that places blame on coal-fired generating plants. Particularly useful for the understanding of statistical analyses of acid rain data.

368. Harcourt, S.A. and J.F. Farrar. "Some Effects of Simulated Acid Rain on the Growth of Barley and Radish," Environmental Pollution, Vol. 22, 1980, pp. 69-73.

Technical report of laboratory study. Results are mixed and not particularly pathological.

369. Harr, T.E. and P.E. Coffey. Acid Precipitation in New York State, Technical Paper No. 43. Albany, NY: New York State Department of Environmental Conservation, 1975.

A general review of acid rain effects, primarily in the Adirondacks. Data presented are not extremely technical but are useful for comparison with later information.

370. Harrison, P. "Population, Climate and Future Food Supply," Ambio, Vol. 13, no. 3, 1984, pp. 161-167.

A good scientific article for a general scientific audience. Focuses on air pollution problems which affect food production. Specific focus is on CO_2 levels but other combustion effluents are discussed peripherally. Effects rather than cause or process oriented.

371. Harrison, R.M. "Important Air Pollutants in the Aquatic Environment," in R.M. Harrison (ed), Pollution: Causes, Effects and Control. London: Royal Society of Chemistry, 1983, pp. 156-175.

A general overview of anthropogenic chemicals in the air; treats SO_2 and NO_2 specifically as critical problems. Also provides a general synopsis of the chemical and physical processes involved in the production of pollutants. Not too technical for non-scientists.

372. Harvey, H.H., et al. Acidification in the Canadian Environ-

ment: Scientific Criterion for an Assessment of the Effects of Acidic Deposition on Aquatic Ecosystems, NRC Report no. 18475. Ottawa: National Research Council of Canada, 1981.

A good scientific policy paper filled with research reports and synopses of research. Meant to be background for political action.

373. Haskell, E.H. The Politics of Clean Air. New York: Praeger Publishers, 1982.

A thorough case study of the development of emission standards and pollution control levels for coal-fired power plants by the U.S. E.P.A. Covers the time period 1977-1979. Primarily for policy scientists.

374. Haynie, F.H. and J.P. Upham. "Effects of Atmospheric Sulphur Dioxide on the Corrosion of Zinc," Materials Protection and Performance, Vol. 9, 1970, p. 35.

An early study of the impact of acid rain on materials. Useful in sequence with later work on dry deposition on metals.

375. Heagle, A.S. and W.S. Heck. Field Methods to Assess Crop Losses Due to Oxidant Air Pollutants, Miscellaneous Publication No. 7, Agricultural Experiment Station, University of Minnesota, St. Paul, Minnesota, 1980.

A how-to manual for agronomists interested in assessment of crop damage caused by acid rain. Results are very general and, in light of what is known now, not always reliable.

376. Heagle, A.S., et al. "Response of Soybeans to Simulated Acid Rain in the Field," Journal of Environmental Quality, Vol. 12, no. 4, 1983, pp. 538-543.

A thorough research report on field research on soybeans using simulated acid rain. This is useful methodologically because of the novel approach, simulating rain in the field conditions, and provides informative and useful data on the overall effects acid rain can have on these plants. Good correlations established between acidity and plant reactions.

377. Heck, W.W., et al. "Air Pollution: Impact on Plants," Proceedings of the Soil Conservation Society of America (pub. no. 132-193-202), 1978.

A general statement of known and suspected acid rain effects on agricultural and forest resources. Particularly interesting because much of the inferential information can now be compared with experimental results.

378. Helvey, J.D., et al. "Acid Precipitation: A Review," Journal of Soil and Water Conservation, Vol. 37, May/June, 1982, pp. 14-16.

A synopsis of research and information available at the time. Very general but a good source of a synopsis of research. Also looks at some of the data shortcomings and areas where research is needed.

379. Helvey, J.D. and J.H. Patric. "Sampling Accuracy of Pit vs Standard Rain Gauges on the Fernow Experimental Forest," Water Resources Bulletin, Vol. 19, no. 1, 1983, pp. 87-89.

A good report on comparative sampling of precipitation for chemical purposes using two "standard" forms of precipitation gathering. Of methodological importance, especially in light of the current controversy over the quality of samples in acid rain research.

380. Helvey, J.D. and J.N. Kochenderfer. "Effects of Acid Precipitation on Nutrient Cycling and Weathering of Minerals in the Central Appalachians," National Acid Precipitation Assessment Program Effects: Research Review, Vol. 2, no. 3, 1983, pp. 47-54.

A concise report on research into soil mineral cycles under acid rain conditions. Particularly useful information on heavy metals; the Appalachians are a critical region of acid rain research at present and these data will begin to fill many gaps in understanding the overall ecosystem effects of acid rain.

381. Henderson, G.S. and W.T. Swank. "Atmospheric Input of Some Cations and Anions to Forest Ecosystems in North Carolina and Tennessee," Water Resources Research, Vol. 12, no. 3, 1976, pp. 24-29.

A rather early report on polluted precipitation in the southeast and its effect on forest soil chemistry. Good baseline data that can be very useful in comparative study to determine trends in ionic activity.

382. Hendrey, G.R., et al. "Acid Precipitation: Some Hydrobiological Changes," Ambio, Vol. 5, no. 5-6, 1976, pp. 224-227.

Presents data on fish community changes in acidified waters but also information on algae, zooplankton and other organic populations lower in the food chain. Acid induced changes in all levels of organisms discussed. Not overly technical.

383. Hendrey, G.R. Acidification of Aquatic Ecosystems: Ecosystem Sensitivity and Biological Consequences, Special Report No. 12. Schenectady, NY: Association for the Protection of the Adirondacks, 1979.

Non-technical report of very technical information on ecosystem reaction to acid rain. Information is useful for regional comparison with other aquatic data. Reports on acidic effects throughout the aquatic food chain.

384. _______, et al. Geological and Hydrochemical Sensitivity of the Eastern United States to Acid Precipitation, EPA-600/3-80-024. Corvallis, OR: U.S. Environmental Protection Agency, 1980.

Technical report of predicted and measured potential impact of acid rain on Eastern U.S. Of particular use to scientists as field testing reveals actual impact. Excellent source of verified regional data.

385. _______. "Response of Freshwater Plants and Invertebrates to Acidification," in Restoration of Lakes and Inland Waters, EPA Report no. 440/5-81-010. Washington, DC: U.S. Environmental Protection Agency, 1980, pp. 457-466.

A comprehensive evaluation of overall effects on subvertebrates and micro- and macroscopic plant life under the influence of acid rain.

386. Hendrey, G.R. "Acid Rain and Gray Snow," Natural History, Vol. 90, No. 2, 1981, pp. 246-255.

General article intended for a general, non-scientific audience. Non-technical, synthetic but a good synopsis of known information.

387. _______ (ed). Early Biotic Responses to Advancing Lake Acidification. Boston: Butterworth Publishers, 1984.

A good collection of articles on ecosystem responses to lacustrine pollution by acid rain. The articles illuminate all portions of the ecosystem that react to pH changes and offer predictions of ultimate resolution of these changes. This collection looks at all levels of ecosystem, from microorganismic to overall lacustine acidification. A very useful and informative book.

388. Henmi, J. "Long-range Transport Model of SO_2 and Sulfate and Its Application to the Eastern United States," Journal of Geophysical Research, Vol. 85, 1980, pp. 4436-4442.

Technical mathematical model for atmospheric physics involved with movement of sulfur combustion compounds through the atmospheric circulation. Useful for regional and national studies.

389. Henriksen, A. "A Simple Approach for Identifying and Measuring Acidification of Freshwater," Nature, Vol. 278, 1979, pp. 542-545.

A technical report on research methodology. Useful to scientists or potential researchers only.

390. ________. "Acidification of Freshwaters--a Large Scale Titration," in D. Drablow and A. Tollan, (eds), Proceedings of the International Conference on Ecological Impacts of Acid Precipitation. Sandefjord, Norway: SNSF, 1980. pp. 68-74.

A very technical report on freshwater chemistry under attack from acid rain. Uses European examples but can be useful for comparative or inferential study. Often referred to as an international standard set of data with which to compare other regional data.

391. Herman, F.A. and E. Gorham. "Total Mineral Material, Acidity, Sulfur and Nitrogen in Rain and Snow at Kentville, Nova Scotia," Tellus, Vol. 9, 1957, pp. 180-183.

A useful set of data on precipitation chemistry in northeastern Canada that can be used for comparative study on a regional or temporal basis.

392. Herrman, R. and J. Baron. "Aluminum Mobilization in Acid Stream Environments, Great Smokey Mountains National Park, U.S.A.," in D. Drablos and A. Tollan (eds), Ecological Impacts of Acid Precipitation. Sandefjord, Norway: SNSF Project, 1980, pp. 218-219.

A short research report on heavy metal reaction to acid rain in a stream environment. One of the first reports on this most important secondary set of effects created by acid rain and made even more important because of the pristine environment in which this was found. Very useful for general ecologists.

393. Hertzberg, P.J. "The Interstate Carriage of Pollutants: The Legal Problem and Existing Solutions," in T.Y. Toribara, et al. (eds), Polluted Rain. New York: Plenum Press, 1980, pp. 463-475.

Provides a Sierra Club perspective on regulation of airborne pollutants across state borders. Primarily an indictment of EPA for giving in to political pressure in relaxing standards with regard to combustion effluent. More an advocacy piece than analytical, but useful for policy students.

394. Hicks, B.B. "On the Dry Deposition of Acid Particles to Natural Surfaces," in T.Y. Toribara, et al. (eds), Polluted Rain. New York: Plenum Press, 1980, pp. 327-339.

A rather short general presentation on the process of dry deposition of acidic materials primarily on leaf surfaces. While the assumed process is discussed, more questions are raised than are answered.

395. _______, et al. Critique of Methods to Measure Dry Deposition, EPA 600/9-80-050. Research Triangle Park, NC: U.S. E.P.A., 1980.

A report of a workshop held by EPA to discuss problems and solutions involved with the measurement of acidic dry deposition. A good methodological analysis but does not provide very many positive suggestions about ways to go about collecting dry materials.

396. _______ (ed). Deposition Both Wet and Dry. Boston: Butterworth Publishers, 1983.

A collection of technical reports on the deposition of air-borne pollutants. Particularly useful because it treats a wide range of pollutants, not just acid rain. Also gives some insight into level of dry deposition, which is a lesser phenomenon.

397. Hidy, G.M., et al. International Aspects of the Long Range Transport of Air Pollutants, ERT #P-5252. Washington, DC: U.S. Department of State, 1978.

Presents several models of atmospheric mixing and transport of combustion effluent materials. Focuses primarily on U.S.-Canada examples but of use for comparative purposes in other regions. Good source of data.

398. Hidy, G.M. and P.K. Mueller. "Monitoring Airborne Contaminants," in T.Y. Toribara, et al. (eds), Polluted Rain. New York: Plenum Press, 1980, pp. 407-434.

An examination of the processes of monitoring air quality in the U.S. Provides an excellent overview of the methods of collection, evaluation, and data presentation utilized by the major precipitation monitoring programs. Also gives some useful graphical presentations of precipitation data. A very useful article for anyone wanting to collect precipitation in the field.

399. Highton, N.H. and M.G. Webb. "On the Economics of Pollution Control for Sulphur Dioxide Emissions," Energy Economics, Vol. 3, no. 2, 1981, pp. 91-101.

Relatively non-technical presentation of some of the micro and macro economic effects of increased pollution control related to acid rain.

400. Hileman, B. "Acid Precipitation," Environmental Science and Technology, Vol. 15, no. 10, 1981, pp. 1119-1124.

An almost journalistic overview of the acid rain phenomenon. Some general scientific knowledge in a synthetic form but no new information.

401. _______. "Acid Deposition," Environmental Science and Technology, Vol. 16, no. 6, 1982, pp. 323A-327A.

A general article on acid rain. Not particularly in-depth or comprehensive.

402. _______. "Crop Losses from Air Pollutants," Environmental Science and Technology, Vol. 16, no. 9, 1982, pp. 495A-499A.

A general statement of regional and national crop losses based on experiments and predictions carried out by EPA. Seminal article for policy development and analysis.

403. _______. "Acid Rain Perspectives: A Tale of Two Countries," Environmental Science and Technology, Vol. 18, no. 11, 1984, pp. 341A-344A.

A comprehensive case study of Canadian and U.S. acid rain problems. This article looks at the sources, emission levels, and remedial/preventive actions taken, or not taken, by both countries. The Canadian case is really given as an example of what the United States should be doing to solve or prevent the acid rain problems generated by or experienced by the United States. Good article for comparative policy students.

404. Hill, A.C. "Vegetation: A Sink for Atmospheric Pollutants," Journal of the Air Pollution Control Association, Vol. 21, 1971, pp. 342-346.

A general discussion of the ways in which all kinds of atmospheric pollutants find their way into vegetative zones. Especially focused on ozone but also includes sulfur and nitrogen compounds in discussion. Good source of a general, global-scaled analysis of one process underlying acid rain effects.

405. Hill, F.B. and R.F. Adamowicz. "A Model for Rain Composition and the Washout of Sulfur Dioxide," Atmospheric Environment, Vol. 11, 1977, pp. 912-927.

Look at the chemical and physical processes active in the production of one form of acid in acid rain. Technical but useful for an informed general reader.

406. Hilst, G. Plume Model Validation and Development: Field Measurements, Plain Sites, EA-3064. Palo Alto, CA: Electric Power Research Institute, 1984.

A very technical report on field testing of deposition patterns which have been predicted by plume modeling with accepted instruments. A good source of effects data and a very important monograph for anyone wishing to understand the transport and deposition of combustion effluent.

407. Hilst, G. Effects of Humidity and Temperature on Conversion of SO_2 to Particulate Sulfate and Sulfite, EA-3310. Palo Also, CA: Electric Power Research Institute, 1984.

A very technical report on research in atmospheric physics and chemistry on the conversion of combustion pollutants into acid compounds. This looks at the processes which are the root of the acid rain problem with a very scientific, unbiased approach. This study has implications for predicting the amount of acidic fallout that can be expected from certain kinds of meteorological conditions and will be quite useful to plume and depositional modelers.

408. Hindawi, I.J. Air Pollution Injury to Vegetation. Washington, DC: U.S. Government Printing Office, 1970.

A well documented, colorfully illustrated book showing various kinds of pollution that injure plants. Meant for an informed public.

409. Hindawi, I.J., et al. "Response of Bush Bean Exposed to Acid Mist," American Journal of Botany, Vol. 67, 1980, pp. 168-172.

A concise research report of an experiment with simulated acid rain. Important information for agronomists and plant pathologists. The direct effects are not found to be particularly harmful; not enough temporal depth to the study to really document indirect effects.

410. Hofken, K.D. and G. Gravenhorst. "Deposition of Atmospheric Aerosol Particles to Beech and Spruce Forest," in H.W. Georgii and J. Pankrath (eds), Deposition of Atmospheric Pollutants. Boston: Reidel Publishing, 1982, pp. 191-194.

Report of research on the dry deposition of atmospheric pollutants onto leaf and needle surfaces in the Solling forest. Wet deposition and, more importantly, throughfall, was also monitored at this site. Good sets of useful data on the three processes.

411. Hogsett, E.W., et al. "Growth Responses in Spinach to Sequential and Simultaneous Exposure to NO_2 and SO_2," Journal of the American Society of Horticulture Science, Vol. 109, no. 2, 1984, pp. 252-256.

A basic scientific report of laboratory research exploring the effects of atmospheric pollutants on an important agricultural crop. The bottom line is that there is no significant effect. Good comparable data source.

412. Hogstrom, U. "Wet Fallout of Sulfurous Pollutants Emitted From a City During Rain or Snow," Atmospheric Environment, Vol. 8, 1974, pp. 1291-1303.

Rather technical but very complete analysis of local production and deposition of acid precipitation. Excellent analysis of the pre-tall stack pollution situation.

413. Holden, A.V. "Surface Waters and Aquatic Ecology," in Ecological Effects of Acid Precipitation, a report of a workshop held at Gatehouse-of-Fleet, Galloway, U.K., September 4-7, 1978.

An early report of preliminary study of lacustrine ecology under the influence of acid rain. Provides benchmark data for comparison with later results as well as analysis that has held up under new data.

414. Holmes, J. (ed). Energy, Environment, Productivity. Washington, DC: U.S. Government Printing Office, 1973.

A collection of articles only a few of which have any interest in acid rain. The book is a compendium of information about all aspects of industry/environment interrelatedness.

415. Hornbeck, J.W. "Acid Rain: Facts and Fallacies," Journal of Forestry, Vol. 79, no. 7, 1981, pp. 438-443.

A general overview of the science of acid rain research at this point in time. Good material for comparative study.

416. ________, et al. "Seasonal Patterns in Acidity of Precipitation and Their Implications for Forest Stream Ecosystems," in Proceedings of the First International Symposium on Acid Precipitation and the Forest Ecosystem. Washington, DC: Conservation Society, 1976, pp. 597-609.

Comprehensive examination of seasonal nature of acid rain intensity. Analysis of seasonal reactions of streams to precipitation pH is explored thoroughly.

417. Horvath, L. "On the Vertical Flux of Gaseous Ammonia Above Water and Soil Surfaces," in H.W. Georgii and J. Pankrath (eds), Deposition of Atmospheric Pollutants. Boston: Reidel Publishing, 1982, pp. 17-22.

Report on research in Hungary on the flux of atmospheric ammonia over both water and uncultivated soil surfaces. Nitrogen compounds are important in the acid rain process and this report examines the chemical dynamics of one of these nitrogen compounds near the surface. Rather technical for non-chemists.

418. Hovland, J., et al. "Effects of Artificial Acid Rain on Decomposition of Spruce Needles and on Mobilization and Leaching of Elements," Plant and Soil, Vol. 56, No. 3, 1980, pp. 45-51.

Scientific report on experiments in humus formation and mineral/nutrient movement through forest soils in a taiga environment. Particularly useful to soil scientists and forestry scientists considering the economic importance of this type forest.

419. Howard, R. and M. Perley. Acid Rain: The North American Forest. Toronto: House of Anansi Press, 1980.

A documentation of the idea that the problem of acid rain was known of for many years before it was given any attention by Canadian or U.S. politicians. Presents the scientific evidence of acid rain effects along with the evolution of policy as these effects are made public. For a general or policy-oriented audience.

420. Howells, G. "Interpretation of Aquatic pH Trends," Transactions of the American Fisheries Society, Vol. 111, no. 6, 1982, pp. 779-781.

A relatively technical statistical evaluation of data collected from freshwater samples across the United States. Most useful for methodological comparisons.

421. Huckabee, J. (ed). Workshop Proceedings: Effects of Trace Elements on Aquatic Ecosystems. Palo Alto, CA: Electric Power Research Institute, 1984.

A collection of very scientific articles on various aspects of the effects of selected combustion effluent by-products on lakes and streams. Not limited to acid rain or acidic materials but very useful to ecologists and limnologists.

422. Hudy, M., et al. Fish Community Structure and Trace Metal Concentrations in Potentially Acid Sensitive Streams of the Southern Blue Ridge Province. Athens, GA: Georgia Cooperative Fishery Research Unit, 1983.

A thorough technical report on the relationships between fish speciation and fish populations in harmful metal concentrations in mountain streams. The potential for acid-induced changes is examined and the effects inferred for this region. A very useful set of data and analysis for ecologists and fishery scientists.

423. Hughes, P.R., et al. "Effects of Air Pollutants on Plant/Insect Interactions," Environmental Entomology, Vol. 10, 1981, pp. 741-744.

Not specifically focused on acid rain but involved with a sub-system of the food chain that is affected by acid rain and its associated atmospheric phenomena. Rather specialized.

424. _______. "Effects of Air Pollution on Plant-Insect Interactions: Increased Susceptibility of Greenhouse-Grown Soybeans to the Mexican Bean Beetle After Plant Exposure to SO_2," Environmental Entomology, Vol. 11, 1982, pp. 173-176.

Evaluates laboratory experiments with atmospheric pollution. SO_2 deposition, either in dry or wet form, is focus of pollution. Data on direct effects as well as biomass are reported thoroughly.

425. _______. "Increased Success of the Mexican Bean Beetle on Field-Grown Soybeans Exposed to Sulfur Dioxide," Journal of Environmental Quality, Vol. 12, no. 4, 1983, pp. 230-244.

A follow-up report on research first presented in the article above. Provides more complete data and a longer period of analysis.

426. Hultberg, H. and A. Wenblad. "Acid Groundwater in Southern Sweden," in D. Drablos and E.T. Tollan (eds), Ecological Impact of Acid Precipitation. Sandefjord, Norway: SNSF Project, 1980, p. 220.

A short report on pH trends in groundwater in Sweden. This is one of the first examples of intense percolation of acid compounds into the water table. Very useful data for comparative study.

427. Huntzicker, J.J., et al. "Aerosol Sulfur Episodes in St. Louis, Missouri," Environmental Science and Technology, Vol. 18, no. 12, 1984, pp. 962-967.

A thorough report on an experiment in measuring SO_2 for eight days. Identifies two types of episode or rises in aerosol sulfur. One is a local phenomenon, the other regional. Good source of hour-by-hour data as well as analysis.

428. Husar, R.B. On Acid Deposition and Possible Remedies, in-house research report, Washington University, St. Louis, Mo., 1983.

A general but insightful examination of methods designed to reduce the level of effect created by acid rain. Part of the comprehensive plume/transport/deposition study in St. Louis.

429. _______ and J.M. Holloway. Sulfur and Nitrogen Over North America, Ecological Effects of Acid Deposition, SNV PM 1636. Stockholm: National Swedish Environment Protection Board, 1983.

A rather general report which looks at all aspects, from production to effects on ecosystems, of the acid rain phenomenon in North America. No new information but a useful presentation and synthesis of research known at that time.

430. Hutcheson, M.R. and F.P. Hall. "Sulfate Washout From a Coal-fired Power Plant Plume," Atmospheric Environment, Vol. 8, 1974, pp. 23-28.

Report of basic research on chemical effluent from a major sulfur pollution source. Very useful for a scientific understanding of the processes underlying production of acid rain.

431. Hutchinson, T.C. and M. Havas (eds). Effects of Acid Precipitation on Terrestrial Ecosystems. New York: Plenum Press, 1980.

An excellent collection of articles on a wide variety of acid rain situations. Most are for a scientific audience but the book is also useful for informed lay readers.

432. Institute for Mining and Minerals Research. Acid Rain in Kentucky. Lexington, KY: University of Kentucky, 1983.

Proceedings of a workshop on acid deposition in Kentucky. Articles contained within include coal technology-related research as well as effects studies. Good source of regional specific data and analysis.

433. Interagency Task Force on Acid Precipitation. National Acid Precipitation Assessment Plan, report to U.S. Environmental Protection Agency. Washington, DC: U.S. Environmental Protection Agency, 1982.

Bureaucratic assessment document useful for an understanding of national policy development and interagency dynamics.

434. Irving, P.M. Response of Field-grown Soybeans to Acid Precipitation Alone and in Combination With Sulfur Dioxide, Ph.D. thesis, University of Wisconsin-Milwaukee, 1979.

Excellent source of data and some analysis of pioneering research. Basic to an understanding of the effects of acid rain on agricultural systems.

435. ________ and J.E. Miller. "Productivity of Field-Grown Soybeans Exposed to Acid Rain and Sulfur Dioxide Alone and in Combination," Journal of Environmental Quality, Vol. 10, 1981, pp. 473-478.

An outgrowth of Ph.D. research reported above but more succinct. No data sets as in dissertation but better analysis and presentation of information. Very useful article.

436. Irving, P.M. "Acidic Precipitation Effects on Crops: A Review and Analysis of Research," Journal of Environmental Quality, Vol. 12, no. 4, 1983, pp. 442-453.

A very thorough review of research results on this topic. Offers data from many projects put into tabular form. Particular focus on soybeans and soybean productivity under acid rain effects.

437. Jacobsen, J.S. and P. Van Leuken. "Effects of Acidic Precipitation on Vegetation," in S. Kasuga, et al. (eds), Proceedings of the 4th International Clean Air Congress. Tokyo: Japanese Union of Air Pollution Prevention Associations, 1977, pp. 96-101.

A very general article, more inferential than factual. No new information but interesting presentations of what was known and suspected at that time.

438. Jacobsen, J.S. "Experimental Studies on the Phytotoxicity of Acidic Precipitation: The United States Experience," in T.C. Hutchinson and M. Havas (eds). Effects of Acid Precipitation on Terrestrial Ecosystems. New York: Plenum Press, 1980, pp. 151-160.

A review of research performed on this segment of the environment in the U.S. Very useful as a synopsis; good analysis provided also.

439. Jansen, J.J. "A Review of the Issues and Its Uncertainties," in C. Curtis (ed), Before the Rainbow: What We Know About Acid Rain. Washington, DC: Edison Electric Institute, 1980, pp. 87-102.

A good article for outlining some of the problems with the current state of acid rain research but the gist of the article is slanted pro-industry. For policy analysts this is a usable statement on acid rain research from the electric utility industry.

440. ________, et al. Utility Acid Precipitation Study Program: 1982 Annual Summary Report. Washington, DC: Edison Electric Institute, February, 1983.

Annual report of data collection and analysis of precipitation samples in the United States. Good source of data that have some continuity.

441. Jeffries, D.S., et al. "Depression of pH in Lakes and Streams in Central Ontario During Snowmelt," Journal of the Fisheries Research Board of Canada, Vol. 36, 1979, pp. 640-646.

A thorough research report on seasonal pH changes in Canadian waters. The resultant lowered pH is assumed to be a result of, and correlates closely to, melting snow in the spring. A good source of melt model for other case studies.

442. Jeffries, D.S. and W.R. Snyder. "Atmospheric Deposition of Heavy Metals in Central Ontario," Water, Air, and Soil Pollution, Vol. 14, 1981, pp. 133-157.

Good report on heavy metal pollution from atmospheric deposition. While not specifically acid rain, the two phenomena are very closely related in process and in occurrence. Good model of atmospheric transport, extensive sets of data, thorough analysis.

443. Jensen, A.L. "Assessing Environmental Impact on Mass Balance, Carrying Capacity and Growth of Exploited Populations," Environmental Pollution, Vol. 36, no. 2, 1984, pp. 133-145.

Exploited populations here means any community, plant or animal, that is harvested by man. The report uses harvesting population prediction models to assess environmental impact. The specific focus is on fish populations and specific mention is made of acid precipitation. Usable models for foresters, agricultural scientists.

444. Jensen, K.F. "Growth Relationships in Silver Maple Seedlings Fumigated With O_3 and SO_2," Canadian Journal of Forestry Research, Vol. 13, 1983, pp. 298-302.

A thorough report on research of simulated combustion pollution effects on maple trees in northern climates. These results, generally quite negative, are not for acid rain per se, but because of the relationship between SO_2 and acid precipitation, this information is useful.

445. Jernelov, A. "The Effects of Acidity on the Uptake of Mercury in Fish," in T.Y. Toribara, et al. (eds), Polluted Rain. New York: Plenum Press, 1980, pp. 211-222.

Report of studies on the ways in which pH affects the bioaccumulation of mercury in lakes and streams. Examines both ecological and physiological processes using data from research in Swedish lakes. Focuses on processes primarily; very little data presented. Useful for ecologists and limnologists.

446. Jockel, J.T. and A.M. Schwartz. "The Changing Environmental Role of the Canada-United States International Joint

Commission," Environmental Review, Vol. 8, no. 3, 1984, pp. 236-251.

A comprehensive historical examination of this important policy and regulatory advisory group. The focus is on the evolution of international acid rain policy.

447. Johnson, A.H. "Evidence of Acidification of Headwater Streams in the New Jersey Pinelands," Science, Vol. 206, 1979, pp. 383-386.

Good research report on stream pollution in a forest environment. Data presented are not comprehensive but are given in useful form and analysis is excellent.

448. _______. "Acidification of Headwater Streams in the New Jersey Pine Barrens," Journal of Environmental Quality, Vol. 8, 1979, pp. 383-386.

Basically the same article as item 447. Useful for general scientific audience but especially useful for hydrologists and aquatic biologists. A landmark study of a pristine region affected by acid rain and snow.

449. _______, et al. "Recent Changes in Patterns of Tree Growth Rate in the New Jersey Pinelands: A Possible Effect of Acid Rain," Journal of Environmental Quality, Vol. 10, no. 4, 1981, pp. 427-430.

A preliminary research report on field studies in a unique environment. Results are based on statistical comparison of observed growth rate changes and reported acidified precipitation. Good source for ecologists.

450. Johnson, A.H. and T.G. Siccama. "Acid Deposition and Forest Decline," Environmental Science and Technology, Vol. 17, no. 7, 1983, pp. 294A-305A.

A good overall analysis of the forest ecosystem under the influence of acid rain. Looks at several U.S. and Canadian forest environments and systems.

451. Johnson, D.W. and G.S. Henderson. "Sulphate Adsorption and Sulfur Reactions in a Highly Weathered Soil Under a Mixed Deciduous Forest," Soil Science, Vol. 128, 1979, pp. 34-40.

A good scientific research report on important soil chemistry under the effects of acid rain. Basic soil chemistry.

452. Johnson, D.W., et al. "Contributions of Acid Deposition and

Natural Processes to Cation Leaching From Forest Soils: A Review," Journal of the American Pollution Control Association, Vol. 33, no. 11, 1983, pp. 167-172.

A good synopsis of research into soil mineral movement under acidic conditions as induced by acid precipitation. No new research but a good overview of the state-of-the-art on this subject at the time of publication.

453. Johnson, M.B. "The Environmental Costs of Bureaucratic Governance: Theory and Cases," in J. Baden and R.L. Stroup (eds), Bureaucracy vs Environment: The Environmental Costs of Bureaucratic Governance. Ann Arbor, MI: University of Michigan Press, 1981, pp. 217-224.

A philosophical discussion which examined current U.S. regulatory policy and procedures. Very critical of the agency-oriented system developed in this country. Provides gross cost analysis of specific pollution-related cases.

454. Jones, H.C., et al. Investigation of Alleged Air Pollution Effects on Yield of Soybeans in the Vicinity of Shawnee Steam Plant, E-EB-73-3. Chattanooga, TN: Tennessee Valley Authority, 1973.

A very early study of localized acid rain effects on agriculture. No definitive physiological research, primarily statistical analysis of yield data but useful for comparison with later studies. The problem of documenting agricultural effects remains a critical one.

455. Jones, U.S. and E.L. Suarez. "Impact of Atmospheric Sulfur Deposition on Agroecosystems," in D.S. Shriner, et al. (eds), Atmospheric Sulfur Deposition. Ann Arbor, MI: Ann Arbor Scientific Publisher, 1980, pp. 377-396.

A very technical and comprehensive article that can provide a scientific overview of the known and anticipated effects of acid rain on agriculture. Also useful for policy analysis background.

456. Junge, C.E. "The Distribution of Ammonia and Nitrate in Rain Water Over the United States," Transactions of the American Geophysical Union, Vol. 39, No. 2, 1958, pp. 241-248.

Technical, scientific presentation of data on atmospheric chemistry. Useful for analysis of atmospheric transportation as well as for prediction of effects of deposition.

457. ______ and R.T. Werby. "The Concentration of Chloride, Sodium, Potassium, Calcium, and Sulfate in Rain Water Over the United States," Journal of Meteorology, Vol. 15, No. 5, 1958, pp. 417-425.

Compilation of data on atmospheric chemistry. Good source of basic information. Not for non-scientists.

458. Kamlett, K.S., et al. "Acid Rain: An Environmentalist's Perspective," Natural Resources Law Newsletter, Vol. 14, no. 4, 1982, pp. 6-9.

Written for a general audience. More useful for policy analysts than for scientists. No new data or interpretations.

459. Kaplan, E., et al. "Relationships and Implications for Effects of Acidification on Surface Water in the Northeastern United States," Environmental Science and Technology, Vol. 15, 1981, pp. 534-544.

Short, inferential article utilizing general statements of known research to predict systemic relationships.

460. Katz, M. et al. "Symptoms of Injury on Forest Crop Plants," in Effects of Sulphur Dioxide on Vegetation. Ottawa: National Research Council for Canada, 1939, pp. 51-103.

Very useful and very early report on forest reactions to local acid rain input. Good source of data for temporal comparison and for examination in light of environmental changes since the 1930s.

461. Katzenstein, A.W. Understanding Acid Rain. Washington, DC: Edison Electric Institute, 1984.

A general book meant for a non-scientific reader. Basically presents the industrial perspective on acid rain causes, effects and control strategies. Good source for policy students.

462. Keene, W.C., et al. "Organic Acidity in Precipitation From Remote Areas of the World," in U.S. E.P.A., Proceedings: National Symposium on Recent Advances in Pollutant Monitoring of Ambient Air and Stationary Sources, EPA-600/9-84-001. Washington, DC: U.S. Government Printing Office, 1983, pp. 300-310.

A report on sampling of precipitation for organic acids (mainly carboxylic) in relatively non-industrialized regions. A good source of data to compare with industrialized regions and for later baseline studies of these same regions.

463. Keever, G.J. and J.S. Jacobsen. "Simulated Acid Rain Effects on Zinnia as Influenced by Available Nutrients," Journal of the American Society of Horticultural Science, Vol. 108, no. 1, 1983, pp. 80-83.

An excellent research report documenting the effects of acid rain on the soil and, as a result, on the nutrient availability for a specific plant. Very useful data and analyses which can be inferred for other, more economically important crops and plants.

464. _______. "Response of Glycine max to Simulated Rain: Environmental and Morphological Influences on the Foliar Leaching of Rb," Field Crops Research, Vol. 6, 1983, pp. 241-250.

An excellent report on specific research into the reaction of leaf processes in soybeans to acid applications. Along with the following article, this provides significant insight into the physiological reactions of soybean plants to acid rain.

465. _______. "Response of Glycine max to Simulated Rain: Localization of Injury and Growth Response," Field Crops Research, vol. 6, 1983, pp. 251-259.

A technical but well-written article which focuses more closely on the effects of acid rain on soybeans. This examines the precise mechanisms by which acidity influences the growth patterns of soybean plants. Important information and analysis for agricultural scientists.

466. Kelly, J.M. Atmospheric Contributions to Sulfur Flux in Three Southeastern Deciduous Forests. Muscle Shoals, AL: Tennessee Valley Authority, 1982.

A thorough examination of the dynamics of the sulfur cycle in deciduous forests known to be under the effects of acid precipitation. Excellent source of research methodology, data sets and analysis.

467. Kelly, T. "How Many More Lakes Have to Die?" Canada Today, Vol. 12, no. 2, 1981, pp. 3-6.

An environmentalist's plea for increased control of air pollution. For public consumption; non scientific. Does provide some insight into public reaction to reports of acid rain damage.

468. Kelso, J.R.M. and C.K. Minns. "Current Status of Lake Acidification and Its Effect on the Fishery Resources of Canada," in R.E. Johnson (ed), Acid Rain Fisheries. Bethesda, MD: American Fisheries Society, 1982, pp. 69-82.

A comprehensive, if often too general, report on lacustrine acidification in Canada. Data on fish populations, fish mortality, and fish susceptibility to contamination are primarily from secondary sources. A good overview for ecologists and for economists.

469. Kelso, J.R.M. and J.M. Gunn. "Responses of Fish Communities

to Acidic Waters in Ontario," in G.R. Hendrey (ed), Early Biotic Responses to Advancing Lake Acidification. Boston: Butterworth Publishers, 1984, pp. 105-115.

A thorough examination of the symptoms and responses of various fish communities observed in lakes in that portion of Ontario under the influence of the Sudbury smelters. Provides mortality and population data not only for gross fish community but for specific communities within lakes. Very clearly delineates fish species which are tolerant and those which are susceptible to only minor acidity. Also provides information on the effects of acidity on heavy metal concentrations in various species of freshwater fish. Very useful article for scientist and non-scientist alike.

470. Kenlan, K.H., et al. "Aquatic Macrophytes and pH as Controls of Diversity for Littoral Cladocerans," in G.R. Hendrey (ed), Early Biotic Responses to Advancing Lake Acidification. Boston: Butterworth Publishers, 1984, pp. 63-84.

An examination of chydorid populations in lakes in rural Maine to determine the changes in population structure with lacustine acidification. Results indicate that pH rather than physical habitat influence littoral chydorid populations, smaller and less diverse populations are found as lake acidity increases. Good source of data for a general ecological student and a comprehensive examination for students of macrophytes.

471. Kennedy, V.C. Research and Monitoring of Precipitation Chemistry in the United States. Reston, VA: U.S. Geological Survey, 1977.

An early overview of regional and national precipitation monitoring, atmospheric chemistry and physics, and acid rain effects research in the United States. Actually intended to be groundwork for requests for governmental support of more research and is thus useful for policy analysts as well as general scientists.

472. Kerr, R.A. "Pollution of the Arctic Atmosphere Confirmed," Science, Vol. 212, 1981, pp. 1013-1014.

A short report of research projects in the polar regions. Data are very general and incomplete.

473. ________. "There is More to Acid Rain Than Rain," Science, Vol. 211, 1981, pp. 692-693.

A short and very general statement about the chemical makeup of acid rain. No new data but a concise presentation of "old" data in a form that can prove useful.

474. ________. "Tracing Sources of Acid Rain Causes Big Stir," Science, Vol. 215, 1982, p. 881.

A short note on the controversies created by industries which view themselves as being unfairly singled out for blame in the acid rain investigations. Useful note for policy analysts.

475. Klarer, C.I., et al. "Effects of Sulphur Dioxide and Nitrogen Dioxide on Vegetative Growth of Soybeans," Phytopathology, Vol. 74, no. 9, 1984, pp. 1104-1106.

Report of controlled research on soybean reactions to pre-acidic pollutants. Leaf stippling, which will interfere with photosynthesis, is the only visible effect but an overall reduction in plant weight (biomass) is also reported. The latter effect related linearly with pollutant dosage.

476. Klein, A.E. "Acid Rain in the United States," Science Teacher, Vol. 41, no. 5, 1974, pp. 36-38.

A very general article; an overview for a lay audience. Some technical concepts reduced to a very common level of understanding.

477. Kleinman, L.J., et al. "Time Dependence on Average Regional Sulfur Oxide Concentrations," Proceedings of the Second Conference on Applied Air Pollution Meteorology. Washington, DC: American Meteorological Society, 1980, pp. 166-171.

A rather technical analysis of the temporal variable in atmospheric circulation modeling. Very useful to chemists.

478. Kneese, A.V. and R.C. d'Arge. "Legal, Ethical, Economic and Political Aspects of Transfrontier Pollution," in T.D. Crocker (ed), Economic Perspectives on Acid Deposition Control. Boston: Butterworth Publishers, 1984, pp. 123-134.

The most comprehensive, concise discussion of this subject listed in this collection. Covers the subject in depth but is written for a wide audience. Very useful for policy discussion.

479. Knudsen, D.A. and D.G. Streets. "Mitigation of Acid Rain--Policy Alternatives," Environmental Progress, Vol. 1, no. 2, 1982, pp. 146-153.

Not so much an analysis of policy as a shopping list of potential policies and procedures. Authors show insight into local, regional, national, and international political and legal structures. Very good point of departure for policy discussions.

480. Koerner, R.M. and D. Fisher. "Acid Snow in the Canadian High Arctic," Nature, Vol. 295, 1982, pp. 137-140.

Good presentation of evidence for acidic precipitation in a remote part of the world. Useful for atmospheric modeling and for global effects study. This includes some of the most complete pH data on snow ever gathered.

481. Kohut, R.J., et al. "An Open-Top Field Chamber Study to Evaluate the Effects of Air Pollutants on Soybean Yield," Proceedings, Fourth Joint Conference on Remote Sensing of Environmental Pollution. Washington, DC: American Chemical Society, 1977, pp. 71-73.

A methodological presentation for field studies. Very useful for prospective field workers in biology, agronomy, geography, and geology.

482. ________. "The National Crop Loss Assessment Network: A Summary of Field Studies," Proceedings, 75th Annual APCA Meeting. Washington, DC: APCA, 1982, paper no. 82-69.5.

A report on the NCLAN program as of 1982. Does not report actual field data but does outline the overall research thrust of this national program. Useful for research policy analysts.

483. Koide, M. and E.D. Goldberg. "Atmospheric Sulfur and Fossil Fuel Combustion," Journal of Geophysical Research, Vol. 76, 1971, pp. 6589-6595.

An early analysis of combustion effluent and one of the studies that led to development of the tall stacks policy. Useful treatment of physical and chemical processes in the atmosphere along with data.

484. Krug, E.C. and C.R. Frink. "Acid Rain on Acid Soil: A New Perspective," Science, Vol. 221, 1983, pp. 520-525.

A general scientific look at the ways in which local geology, especially overdurdened soil, interact with acid rain. Also brings in land-use and local vegetation as variables.

485. Kucera, V. "Effects of Sulfur Dioxide and Acid Precipitation on Metals and Anti-rust Painted Steel," Ambio, Vol. 5, 1976, pp. 243-248.

Report on studies of various metal surfaces, both coated and uncoated, exposed to acid rain. Data on corrosion and oxidation rates for all materials and surfaces along with analysis and predictions are presented.

486. Kullenberg, G.E.B. (ed). Pollutant Transfer and Transport in the Sea. Boca Raton, FL: CRC Press, 1982.

Not limited to articles/chapters on acid rain or combustion-produced pollution but it does offer several chapters which deal with acidification of the oceans, combustion-produced pollution in the oceans, or transport and sink processes which are directly applicable to the acid rain processes. Rather technical articles.

487. Kurtz, J. and W.A. Scheider. "An Analysis of Acidic Precipitation in South-Central Ontario Using Air Parcel Trajectories," Atmospheric Environment, Vol. 15, no. 7, 1981, pp. 1111-1116.

A thorough report of experiments in tracing pollutant trajectories on an air-mass scale. By beginning with the acid precipitation event, the experiment was able to trace the pollution back to its "octant of origin." Good model for predicting downwind transport and deposition. Very technical, for atmospheric chemists and physicists.

488. Kyle, P., et al. "The Volcanic Record of Antarctic Ice Cores: Preliminary Results and Potential for Future Investigations," Annals of Glaciology, Vol. 3, 1982, pp. 172-177.

A technical report on analysis of chemical depositions in polar regions. Gives baseline data and proposes methodology for comparative study. Good information for scientists of all disciplines.

489. LaBastille, A. "Acid Rain: How Great a Menace," National Geographic, Vol. 1160, no. 5, 1982, pp. 652-681.

A general article, supported with prolific amounts of pictures and figures, for an informed lay audience. No new information is presented but colorful illustrations of known data are provided.

490. Lang, D.S., et al. "Responses of Plants to Submicron Acid Aerosols," in T.Y. Toribara, et al. (eds), Polluted Rain. New York: Plenum Press, 1980, pp. 273-290.

Report on simulated exposure of vegetation to very small acid aerosol particles. Both bean plants and poplar tree leaves were exposed and the effects monitored. Both acute, short-term negative effects and chronic, long-term effects are reported and analyzed. Good discussion of both methodology and of observed physiological damage.

491. Last, F.T., et al. "Acid Precipitation--Progress and Problems," in D. Drablos and A. Tollan (eds), Ecological Impact of Acid Precipitation. Oslo: SNSF Project, 1980, pp. 10-12.

A journalistic update of scientific and policy initiatives regarding acid rain. Does provide a very general statement of scientific and public perceptions of the phenomenon.

492. Laurence, J.A. and K.L. Reynolds. "Effects of Concentration of SO_2 and Other Characteristics of Exposure on the Development of Xanthomonas phaseoli Lesions in Red Kidney Beans," Phytopathology, Vol. 15, 1982, pp. 345-356.

More research into kidney beans that can have application to other agricultural plants. Of interest to all readers but slightly technical. Good source of basic scientific data.

493. Laurence, J.A. and A.L. Aluisio. "Effects of Sulfur Dioxide on Expansion of Lesions Caused by Corynebacterium nebraskense in Maize and Xanthomonas phaseoli vs. sojensis in Soybean," Phytopathology, Vol. 71, 1981, pp. 445-448.

Basic scientific research on the ways in which acid rain creates opportunities for viral and bacteriological infestations on agricultural plants. For scientists.

494. Laurence, J.A., et al. "Effect of Sulfur Dioxide on Southern Bean Mosaic and Maize Dwarf Mosaic," Environmental Pollution, Vol. 24, 1981, pp. 185-191.

Report of research into the effects of aerosol pollution on major crop diseases. For a general informed audience of agriculturalists or for background information for scientists.

495. Lauver, T.L. and D.C. McCune. "Kinetics of Removal of Particulate Deposits from Foliage by Precipitation," in J.S. Jacobson and L.S. Raymond (eds), Proceedings of the Second New York Symposium on Atmospheric Deposition. Albany, NY: State University of New York, 1984, pp. 83-90.

A very technical report on the physics and chemistry of dry deposition removal from foliage by subsequent precipitation. Of limited utility to non-scientists.

496. Lazrus, A.L., et al. "Sulfur and Halogen Chemistry of the Stratosphere and of Volcanic Eruption Plumes," Journal of Geophysical Research, Vol. 84, 1979, pp. 7869-7875.

Very technical analysis of upper air chemistry with particular emphasis on volcanic effluent. Of use for delineating natural from anthropogenic acids in the atmosphere.

497. LeBlanc, D.C., et al. "The Use of Stem Analysis Procedures to Study the Impact of Acidic Deposition on Tree Growth," in Proceedings of the Second New York State Symposium on Atmospheric Deposition. Albany, NY: State University of New York Environmental Research Center, 1983, pp. 148-155.

A rather specific research report on experiments with methods with which to measure the effects of acid deposition on plants. For botanists or forestry scientists only but of importance to that group.

498. Lee, J.J. and D.E. Weber. "The Effect of Simulated Acid Rain on Seedling Emergence and Growth of Eleven Woody Species," Forest Science, Vol. 25, 1979, pp. 393-398.

Good basic research on acid rain effects in a forest environment. One of the first studies of forest growth under acidic conditions. Good source of basic research data.

499. ________. Effects of Sulfuric Acid Rain on Two Model Hardwood Forests, EPA 600/3-80-014. Corvallis, OR: Corvallis Environmental Research Laboratory, 1980.

Comprehensive report of simulated and real acid rain in a forest environment. Good source of data and some analysis.

500. Lee, J.J., et al. "Effect of Simulated Sulfuric Acid Rain on Yield, Growth, and Foliar Injury of Several Crops," Environmental and Experimental Botany, Vol. 21, 1981, pp. 175-185.

A comprehensive report of several ongoing experiments with acid rain and agricultural plants. The results are mixed but the information is important for anyone interested in the considerable economic effects of acid rain in the U.S.

501. Leibried, R.T., et al. "Acid Precipitation Runoff and Its Effect on Source Water Quality for Two Small Water Supplies," Journal of the American Water Works Association, Vol. 46, no. 4, 1984, pp. 56-66.

A very useful report on study of the impact of acid precipitation on small water supply reservoir lakes. This is rather novel research because it looks at the water itself as the economic entity rather than the fish or other wildlife which live in it. While the implications are not overly pessimistic, these data can be especially important in small water supply systems.

502. Leivistad, H. and I.P. Muniz. "Fish Kill at Low pH in a Norwegian River," Nature, Vol. 259, 1976, pp. 391-392.

One of the first reports of documented ecological damage in streams from acid rain. Just a research note but baseline data.

503. Leland, E.W. "Nitrogen and Sulfur in the Precipitation at Ithaca, New York," Agronomy Journal, Vol. 44, 1952, pp. 172-175.

Short report of precipitation chemistry in one location in the heart of the most acidic region in the U.S. Very early study which is useful for comparison.

504. Leonard, R.L., et al. "Some Measurements of pH and Chemistry of Precipitation at Davis and Lake Tahoe, California," Water, Air, and Soil Pollution, Vol. 15, 1981, pp. 153-167.

Excellent technical report of water chemistry in California. Acid rain is not a real problem in that region yet but this can become regional baseline data.

505. Levine, J.S., et al. "Simultaneous Measurements of NOS, NO, and NO_3 Production in a Laboratory Discharge: Atmospheric Implications," Geophysical Research Letters, Vol. 8, 1981, pp. 357-360.

Report of basic laboratory research on a process that underlies production or non-production of acidic precipitation. Not for non-scientists.

506. Lewis, R.A., et al. "The Effects of Coal-fired Power Plant Emissions on Vertebrate Animals in Southeastern Montana," in U.S. E.P.A., The Bioenvironmental Impact of a Coal-fired Power Plant, EPA-600/3-78-021. Washington, DC: U.S. Government Printing Office, 1978, pp. 213-279.

A progress report of a study of small mammals and birds near a power plant. The study looks at the structure and dynamics of these populations as well as the physiological, biochemical and behavioral reactions of each species. A landmark study of animal ecology under the effects of acid rain and related phenomena.

507. Lewis, W.M. and M.C. Grant. "Acid Precipitation in the Western United States," Science, Vol. 207, 1980, pp. 176-177.

A short journalistic report of a relative non-phenomenon. Of interest because of the relative delicacy of the western environments.

508. Liddle, G. "Potential Socio-economic Impacts of Acid Rain on the Fishing Lodge Industry of Northern Ontario," Proceedings of the Action Seminar on Acid Precipitation. Toronto: Ontario Ministry of the Interior, 1979.

Early implication report of damage to fishing based on incomplete findings at the time. The projections are too severe but reflect the hysteria of the period.

509. Likens, G.E. The Chemistry of Precipitation in the Central Finger Lakes Region, Technical Report 50. Ithaca, NY: Cornell University Water Resources Center 1972.

One of the first modern acid rain reports for the eastern U.S. Likens establishes his reputation as a premier scientist with this comprehensive data compilation. Basic information for scientific analysis.

510. _______, et al. "Acid Rain," Environment, Vol. 14, no. 2, 1972, pp. 33-40.

An early general article for an informed lay audience. Outlines the implications for acid rain as known at the time. Of use to policy students as well as scientists.

511. _______ and F.H. Bormann. "Acid Rain: A Serious Regional Environmental Problem," Science, Vol. 184, 1974, pp. 1176-1179.

An overview of the implications of acid rain for the United States and Canada based on early research. For an informed lay audience or a general scientific audience looking for a regional article. No new data or analysis.

512. _______, et al. "Hydrogen Ion Input to the Hubbard Brook Experimental Forest, New Hampshire, During the Last Decade," Proceedings of the First International Symposium on Acid Precipitation and the Forest Ecosystem held in Columbus, Ohio. Columbis, OH: Ohio State University, 1975, pp. 38-43.

A landmark report of longitudinal research on a site in one of the most polluted pristine environments in the eastern U.S. The data on forest ecology are very important.

513. _______. "Acid Precipitation," Chemical and Engineering News, Nov. 22, 1976, pp. 29-44.

An overview article touching on physics, chemistry, and ecology. Written technically but for a general scientific audience. A good introduction to early scientific thinking on acid rain.

514. _______, et al. "Acid Rain," Scientific American, Vol. 43, Oct., 1979, p. 241.

A short update on the progress of research into the acid rain phenomenon in the U.S.

515. _______. "Variations in Precipitation and Streamwater Chemistry at the Hubbard Brook Experimental Forest During 1964 to 1977," in T.C. Hutchinson and M. Havas (eds), Effects of Acid Precipitation on Terrestrial Ecosystems. New York: Plenum Press, 1980, pp. 443-464.

An update of previous publications about the longitudinal study on forest ecology in the pristine northeast U.S. Very useful for comparison with other stream basins and quite comprehensive.

516. _______ and T.J. Butler. "Recent Acidification of Precipitation in North America," Atmospheric Environment, Vol. 15, 1981, pp. 1103-1109.

A general article on the atmospheric effects related to acid rain in the U.S. Less meteorology than effects oriented.

517. Liljestrand, H.M. and J.J. Morgan. "Chemical Composition of Acid Precipitation in Pasadena, California," Environmental Science and Technology, Vol. 12, 1978, pp. 1271-1273.

A straightforward data presentation from precipitation monitoring in inland California. Good for regional comparisons; weak on analysis.

518. _______. "Modeling the Chemical Composition of Acid Rain in Southern California," in T.Y. Toribara, et al. (eds), Polluted Rain. New York: Plenum Press, 1980, pp. 109-124.

A report of precipitation analysis in the Pasadena, California region in 1976-1977. This discussion centers on the methods by which acidity is determined and on the acid constituents which lower the precipitation pH. Charge balance and conductivity balance are discussed in this rather technical chemical report.

519. Lindberg, S.E. and R.C. Harriss. "Water and Acid Soluble Metals in Atmospheric Particles," Journal of Geophysical Research, Vol. 88, 1983, pp. 5091-5100.

A rather technical report on the chemical and physical processes involved with incorporation of heavy metals into precipitation. The general results indicate that acidic precipitation is more amenable to uptake of metals and may, indirectly, lead to more metal pollution. This is useful information for effects-oriented researchers also because of the increasing evidence that metal toxicity may be the direct cause of physiological damage to plants in acidic situations.

520. Lindberg, S.E. and G.M. Lovett. "Application of Surrogate Surface and Leaf Extraction Methods to Estimation of Dry Deposition to Plant Canopies," in H.R. Pruppacher, et al. (eds), Precipitation Scavenging, Dry Deposition and Resuspension. New York: Elsevier Science Publishing, pp. 837-848.

A thorough examination of one method of measuring dry deposition of aerosol pollutants in forest situations. This study can have application to agricultural situations also and can be of use to a wide range of scientists.

521. Lindberg, S.E., et al. "Mechanisms of the Flux of Acidic Compounds and Heavy Metals onto Receptors in the Environment," in Proceedings of the Colloquium on Acid Precipitation: Origin and Effects. Lindau, FRG: Berichte, 1983, pp. 165-171.

A very technical report on the chemical and physical processes active as heavy metals move through the atmosphere. This provides even more evidence of the relationship between combustion effluent, acid precipitation and heavy metal transport and deposition. Of interest to chemists and effects scientists.

522. Linthurst, R.A. (ed). Direct and Indirect Effects of Acidic Deposition on Vegetation. Boston: Butterworth Publishers, 1984.

A comprehensive compilation of articles from respected scientists. Not for a lay audience. Contains results of unpublished research as well as synopses of research already available. One of the modern standards in this field.

523. Linzon, S.N., et al. "Sulphur Dioxide Injury to Vegetation in the Vicinity of a Sulphite Pulp and Paper Mill," Water, Air, and Soil Pollution, Vol. 2, 1973, p. 129.

An early study of localized pollution damage. This is especially useful to compare with the post "tall stacks" situation.

524. ________. "Terrestrial Effects of Long Range Pollutants--Crops and Soils," Proceedings of the APCA Conference, April 7-8, 1981. Montreal: APCA, 1981.

A general synthesis of available research. Can provide a quick overview of this subject. Of real utility to soil scientists and agronomists.

525. Lodge, J.P., et al. Chemistry of U.S. Precipitation: Final Report on the National Precipitation Sampling Network. Washington, DC: National Center for Atmospheric Research, 1968.

A compendium of data collected over several years across the U.S. This report was used in development of subsequent air quality regulations and policy.

526. Logan, J. "Nitrogen Oxides in the Troposphere: Global and Regional Budgets," Journal of Geophysical Research, Vol. 88, 1983, pp. 19785-10807.

Examines the level of nitrogen compounds in various types of regions around the world. He focuses somewhat on methodological questions as well as presentation and analysis of the data. Several significant reductions in pollution levels are put forth. Good reading for meteorologists and chemists.

527. Luoma, J.R. "Troubled Skies, Troubled Waters," Audubon, Vol. 82, no. 6, 1979, pp. 43-52.

A journalistic article for general audiences. A shallow overview but of some use as an indicator of public knowledge and concern for the acid rain phenomenon.

528. Lykke, E. "Pollution Problems Across International Boundaries," in T.Y. Toribara, et al. (eds), Polluted Rain. New York: Plenum Press, 1980, pp. 477-488.

A policy article examining the various OECD-like organizations that have been developed in Europe and through the United Nations to deal with the problem of transboundary pollution. Basically a short history of these organizations and, as such, of use to students of public and international policy.

529. Lynch, J.A. and E.S. Corbett. "Acid Precipitation--A Threat to Aquatic Ecosystems," Fisheries, Vol. 5, no. 3, 1980, pp. 8-12.

Primarily a review of published research that deals with aquatic ecosystems and acid rain. Not technical but of use by lay and scientific readers alike.

530. Lynch, J.A. "Acid Precipitation: Detriment or Benefit?" Science in Agriculture, Vol. 28, No, 4, Summer, 1981, p. 7.

A synthetic article using other research as a basis for conjecture. The statements about the good that acid rain can do in forests casts doubt on the validity of other conclusions the author offers.

531. Lyons, W.A. "Evidence of Transport of Hazy Air Masses From Satellite Imagery," Annals, of the New York Academy of Science, Vol. 338, 1980, pp. 418-433.

Well-written report of using remote sensing to track polluted air masses. Especially useful in developing predictive models for downstream effects.

532. MacIntire, W.H. and J.B. Young. "Sulfur, Calcium, Magnesium, and Potassium Content and Reaction of Rainfall at Different Points in Tennessee," Soil Science, Vol. 15, 1923, pp. 205-227.

A very early report on the ionic activity in precipitation both in the air and in the soil. Basic research for background reading.

533. MacLaughlin, S.B. and C.F. Baes. "Trace Elements in Tree Rings: Evidence of Recent and Historical Air Pollution," Science, Vol. 224, no. 4648, 1984, pp. 494-496.

Very useful research report on work done in the Smoky Mountains in and around Copper Hill, Tennessee, a major copper smelter site. Implications for use of tree ring data for pollution monitoring is quite important. Data provided are useful for regional and temporal comparative study. For a general scientific audience.

534. Mahlum, D.D., et al. (eds). Developmental Toxicology of Energy-Related Pollutants. Washington, DC: U.S. Department of Energy, 1978.

The proceedings of the 17th Hanford Biology Symposium. Papers include several which can be of interest to students of combustion effuent and human health but the majority of articles focus on radiation-related phenomena.

535. Maloney, M.T. and B. Yandle. "Bubbles and Efficiency--Cleaner Air at Lower Cost," Regulation, May/June, 1980, pp. 49-52.

Two environmental advocates present a "best case" scenario of air pollution control. Good easy reading for policy development study.

536. Mandelker, D.L. "Emission Quotas for Maintaining Air Quality," in T.Y. Toribara, et al. (eds), Polluted Rain. New York: Plenum Press, 1980, pp. 449-462.

A policy analysis article examining the ways in which the NAAQS are being and can be implemented by EPA. Examines the state-by-state approach to regulation vs the regional approach. Four alternate strategies are discussed.

537. Mandl, R.H., et al. "A Cylindrical Open-top Chamber for Exposure of Plants to Air Pollutants in the Field," Journal of Environmental Quality, Vol. 2, 1973, pp. 371-376.

Methodological report dealing with one of the real problems in acid rain/agriculture research.

538. Manning, R.C. "Air Pollution: Group and Individual Obligations," Environmental Ethics, Vol. 6, no. 3, 1984, pp. 211-226.

An analysis of the ethical and moral attitudes exhibited by institutions and individuals toward the problems and solutions to problems concerning air pollution. Most of the article deals with the actions and reactions of individuals, both as members of society and as single actors. Useful for policy analysis.

539. Maroulis, P.J. and A.R. Bandy. "Estimate of the Contribution of Biologically Produced Dimethyl Sulfide to the Global Sulfur Cycle," Science, Vol. 196, 1977, pp. 647-648.

A short research note inferentially reporting on the rate of naturally occurring sulfur compounds in the upper atmosphere. Not thorough.

540. Marshall, E. "Acid Rain, a Year Later," Science, Vol. 221, 1983, pp. 241-242.

A short journalistic report on progress in acid rain research and policy development. Useful as a chronicle of public attitudes and scientific community reaction and action both as researchers and as citizens.

541. Martens, I.J. Report on the Nickel Industry--Executive Summary. Ottawa: Environment Canada, 1980.

An overview which includes considerable reference to air pollution and acid rain associated with the smelters in Canada.

542. Mayer, R. and B. Ulrich. "Conclusions on the Filtering Action of Forests From Ecosystem Analysis," Ecology, Vol. 9, no. 2, 1974, pp. 157-168.

Provides insight into the way in which forest systems utilize and transform precipitation chemicals. Rather technical for a nonscientific reader.

543. McBean, G.A. "Acid Rain: Emissions, Atmospheric Transport Transformation and Depositions," Proceedings of the Action Seminar on Acid Precipitation. Toronto, Ontario: ASAP Coordinating Committee, 1979, pp. 9-14.

A general look at the international transport problem in North America. A Canadian perspective for a lay audience with an "environmentalist" outlook. No new information or analysis of the problem of acid rain but this can provide some insight into public concerns and thus into the development of public policies.

544. McClenahen, J.R. "The Impact of an Urban-Industrial Area on Deciduous Forest Tree Growth," Journal of Environmental Quality, Vol. 12, 1983, pp. 64-69.

Using tree-ring and other forest data, this article makes a very convincing case against combustion effluent from industrial and urban centers. Because this article examines only one effect of combustion effluent, it provides a more in-depth analysis than do previous general effects-oriented articles and is very useful in understanding the ecological effects of acid rain on a regional and local level.

545. McClenahen, J.R. and N.H. McCarthy. Report on Tree Ring Responses to Site, Climate, and Air Pollutants in the Cuyahoga Valley National Recreation Area. Washington, DC: National Park Service, 1983.

A thorough technical report on tree ring observations in a forest believed to be heavily affected by acidic deposition. These data provide considerable temporal depth to trends in reduced forest health that have been reported from foliar or other external observations. This type of data strengthens the case against acid rain as an anthropogenic phenomenon.

546. McColl. J.G. A Survey of Acid Precipitation in Northern California, Final report of California Agricultural Experiment Station Project CA-B-SPN-3664-H. Davis, CA: U.S. Department of Agriculture, 1980.

Precipitation monitoring report. Useful tables of data that are good for baseline comparison.

547. McDonald, N.W. The Effects of Simulated Acid Precipitation on Regeneration and Soils in the Jack Pine-Grayling Sand Ecosystem, M.S. Thesis. E. Lansing, MI: Michigan State University, 1983.

A thorough report of ecosystem regeneration in an acid rain environment. While this simulation does not provide the temporal depth necessary to fully understand environmental reaction to acid rain it does provide a glimpse of what might happen if rain acidity were reversed.

548. McDowell, T.R. and J.M. Omernik, Non-point Source--Stream Nutrient Relationships: A Nationwide Study, EPA-600/3-79-103, U.S. Environmental Protection Agency. Corvallis, OR: Environmental Research Laboratory, 1979.

Comprehensive compilation of data from a national network. Useful for any stream basin study. Particularly concerned with land-use patterns, more so than with precipitation.

549. McElroy, M.W. Retrofit NO_x and SO_2 Controls for Coal-fired Utility Boilers. Palo Alto, CA: Electric Power Research Institute, 1983.

A thorough technical analysis of the use of new pollution control technology in "old" coal-fired plants to reduce the level of combustion effluent related to acid precipitation. Looks at technological problems as well as the economics of this situation.

550. McFee, W.A. "Effects of Acid Precipitation and Atmospheric Deposition on Soils," in J.N. Galloway, et al. (eds), A National Program for Assessing the Problem of Atmospheric Deposition (Acid Rain). Fort Collins, CO: Natural Resource Ecology Laboratory, 1978, pp. 64-73.

An early analysis of acid rain effects on the soil. More inference than evidence but, as background reading, well done.

551. _______. Sensitivity of Soil Regions to Acid Precipitation, EPA-600/3-80-013, U.S. Environmental Protection Agency. Corvallis, OR: Environmental Research Laboratory, 1980.

Large-scale analysis of acid rain effects on the soil. Utilizes basic geological principles and some research reports for regional statements and analysis of soil processes. A rather technical work by one of the international leaders in soil science.

552. _______. "Effects of Atmospheric Pollutants on Soils," in T.Y. Toribara, et al. (eds). Polluted Rain. New York: Plenum Press, 1980, pp. 307-323.

Examines the effects of both acid precipitation and various metal ions on the nutrient availability in soils. Primarily a theoretical treatise on the processes involved but does discuss reported effects from a variety of sources. Metal pollution is seen as a serious long-term problem.

553. _______. Sensitivity Ratings of Soils to Acid Deposition: A Review. Palo Alto, CA: Electric Power Research Institute, 1982.

A rating of soil types, based on earlier research and writing by the author, to determine their ability to buffer acidified precipitation. While a good report and scientifically accurate, this falls into the category of pro-industry interpretation, placing the blame for acid rain effects on factors other than the electric power industry. Useful data for regional analysis.

554. McLean, R.A.N. "Mercury Transport in the Environment--Analytical and Sampling Problems," in T.Y. Toribara, et al., (eds), Polluted Rain. New York: Plenum Press, pp. 151-174.

A methodological treatise examining the ways in which trace heavy metal concentrations are determined in precipitation and in standing water. Study performed in Ontario and Quebec. Very useful analysis of a problem closely related to acid rain and combustion effluent pollution of lakes and streams.

555. McQuattie, C.J. "Effect of Simulated Acid Rain on Endomy Corrhizal Development and Growth of Black Locust," Phytopathology, Vol. 74, no. 10, 1984, p. 1270.

An abstract of a research report on laboratory testing of acid rain on a woody legume. Rhizobial activity was significantly reduced with precipitation pH down to 2.6 but there was no noticeable plant growth reduction. The implications for reduced future tree growth are clear. Good report.

556. Meinert, D.L. and F.A. Miller. A Review of Water Quality Data in Acid Sensitive Watersheds Within the Tennessee Valley--Volume 1, Tennessee Valley Authority resource paper. Chattanooga, TN: Tennessee Valley Authority, 1981.

Analysis of massive amounts of data with great longevity. While some inferences are made about acidic precipitation, the emphasis is on land-use. Does discuss geological controls of acid rain.

557. Mendelsohn, R. and G. Orcutt. "An Empirical Analysis of Air Pollution Dose-Response Curves," Journal of Environmental Economics and Management, Vol. 6, no. 2, 1979, pp. 85-106.

Very technical econometric analysis of acid rain effects. Not for the non-economist.

558. Menser, H.A. and G.H. Hodges. "Effects of Air Pollutants on Burley Tobacco Cultivation," Agronomy Journal, Vol. 62, 1970, pp. 265-269.

A very early report on overall air pollution effects on one of the most economically important crops in the United States. The results are rather general and not always related directly to acid rain but are quite useful as background information.

559. Mentz, F.C. and C.T. Driscoll. "An Estimate of the Costs of Liming to Neutralize Acidic Adirondack Surface Waters," Water Resources Research, Vol. 19, no. 5, 1983, pp. 1139-1149.

A report of liming experiments in New York state and analysis of those experiments from a cost-effective point of view. Develops an econometric model for estimating liming costs and for estimating overall environmental effects of liming. Useful in a general sense to economists and to students of public policy.

560. Mentz, F.C. and J.K. Muller. "Acidification Impact on Fisheries: Substitution and the Valuation of Recreation Resources," in T.D. Crocker (ed), Economic Perspectives on Acid Deposition Control. Boston: Butterworth Publishers, 1984, pp. 135-156.

Relatively technical report on one set of economic concerns for acid rain effects. The gist is that it is still too soon to prove but not too soon to speculate.

561. Miller, J.E., et al. "Open-Air Fumigation System for Investigating Sulfur Dioxide Effects on Crops," Phytopathology, Vol. 70, 1980, pp. 1124-1128.

Methodological report for aerosol pollution studies by agronomists. Technical, specialized, not for general consumption.

562. Miller, T. Acid Rain Legislation a Serious Threat. New York: Standard and Poors, 1982.

A Wall Street perspective using the idea that any government regulation is bad for business and industry. General, policy oriented.

563. Mills, K.H. "Fish Population Responses to Experimental Acidification of a Small Ontario Lake," in G.R. Hendrey (ed), Early Biotic Responses to Advancing Lake Acidification. Ann Arbor, MI: Ann Arbor Scientific Publishing, 1984, pp. 117-131.

Report of a five-year study in Canada during which lake pH was experimentally lowered each spring to simulate acid snowmelt. Chemical changes in the lakes overall are reported as are fish population and speciation data. A definitive relationship was established between lake pH and fish population. Good data for comparative analysis.

564. Mollitor, A.V. and D.J. Raynal. "Acid Precipitation and Ionic Movements in Adirondack Forests," Soil Science Society Journal, Vol. 46, 1982, pp. 137-141.

Excellent report on research in forest soils. Results are inconclusive about ultimate effects of acid rain on forests but this will provide an important stepping stone to a conclusion.

565. _______. "Atmospheric Deposition and Ionic Input in Adirondack Forests," Journal of the Air Pollution Control Association, Vol. 33, 1983, pp. 1032-1036.

A technical but very clearly and concisely written report on depositional chemistry in a forest environment. This article is important because it examines the depositional process at the ionic rather than grossly chemical level and is a natural progression in acid rain research.

566. Montgomery, T.L., et al. "A Simplified Technique Used to Evaluate Atmospheric Dispersion of Emission from Large Power Plants," in K.E. Noll and W.T. Davis (eds), Power Generation: Air Pollution Monitoring and Control. Ann Arbor, MI: Ann Arbor Science Publishers, 1976, pp. 49-64.

A compilation of various procedures, presented step-by-step, to calculate surface concentrations of chemicals from coal burning sites. Presents a synthetic model which factors in such variables as stack height, wind speed, stack diameter and emission rates. Rather technical but useful for atmospheric scientists and regional studies.

567. Moore, T. "Dry Capture of SO_2," EPRI Journal, Vol. 9, no. 2, 1984, pp. 14-21.

Well-written examination of new mineral reagent technology that can remove 70-80% of the SO_2 from low sulfur coal combustion sources. A useful article that presents some of the research ongoing in preventive technology.

568. ________. "The Retrofit Challenge in NO_x Control," EPRI Journal, Vol. 9, no. 9, 1984, pp. 26-33.

An examination of new technology available for low-NO_x burners which are meant to reduce coal combustion emissions. Very process oriented and specific to this technology but it does also discuss the economics of combustion control.

569. Moskowitz, P.D., et al. Oxidant Air Pollution: Estimated Effects on U.S. Vegetation in 1969 and 1974. Upton, NY: Brookhaven National Laboratory, 1980.

Synthetic presentation utilizing inferential models as much as research evidence. Good general, moderately technical reading.

570. Munger, W.J. and S.J. Eisenreich. "Continental-scale Variations in Precipitation Chemistry," Environmental Science and Technology, Vol. 17, No. 1, 1983, pp. 32A-42A.

A macro analysis of the acid rain phenomenon. Particularly useful for an overview of North American precipitation trends.

571. Munton, D. "Dependence and Interdependence in Transboundary Environmental Relations," International Journal, Vol. 36, no. 1, 1980, pp. 139-184.

A comprehensive look at the political, legal, and philosophical variables active in making or not making international environmental policy decisions. Good for economists and policy students.

572. Murray, F. "Responses of Subterranean Clover and Ryegrass to Sulphur Dioxide Under Field Conditions," Environmental Pollution, Vol. 36, no. 3, 1984, pp. 239-250.

A research report on open-chamber exposure of crops to SO_2. Although some chemical changes were noted in the plants, no reduction in either root or shoot weight were reported. This adds to the literature which indicates little or no direct acid rain effect on many agricultural plants. Good sets of data.

573. Muschett, F.D. "Environmental Pollution: Spatial Variations and Contexts for Growth Management," in J.W. Frazier and B.J. Epstein (eds), Applied Geography Conferences, Vol. 2. Binghamton, NY: State University of New York, 1979, pp. 203-212.

This provides an attempt to predict pollutant discharge from urban regions based on economic and social indicators. While the model might not be universally applicable, the author does provide some insight into the processes that underlie creation of air pollution levels as he develops his model. A useful background article.

574. National Air Pollution Control Administration. Kanawha Valley Air Pollution Study, APTD 70-1. Washington, DC: U.S. Environmental Protection Agency, 1970.

An extensive report, data compilation, and analysis of one of the major coal burning regions in the United States. Provides a complete air pollution monograph which can be used comparatively with other regions or with itself at a later time. Good source for all environmental scientists.

575. National Atmospheric Deposition Program. A National Program for Assessing the Problem of Atmospheric Deposition (Acid Rain), Report NO. NC-141. Fort Collins, CO: Natural Resource Ecology Laboratory, 1978.

Outlines the administrative development of the network of monitoring sites and collection centers of the NADP. Useful in understanding the background of research policy on a national level.

576. _______. NADP Report: Precipitation Chemistry, Volumes I-III, 1978-1980. Fort Collins, CO: Colorado State University, 1981.

Comprehensive data sets for United States precipitation monitoring program. Useful for baseline or for a comparative study using subsequent data sets. This is one of a series of the most comprehensive data sets on precipitation chemistry in the United States.

577. National Coal Association. Some Facts About Acid Rain. Washington, DC: National Coal Association, 1982.

An interesting analysis of the acid rain phenomenon by an organization opposed to any further regulation of combustion effluent.

578. National Economic Research Associates, Inc. A Report on the Results from the Edison Electric Institute Study of the Impacts of the Senate Committee on Environment and Public Works Bill on Acid Rain Legislation, consultant report, Washington, DC, June 1983.

A relatively unbiased analysis of a study conducted by an electric utility policy and lobbying organization (Edison Electric). A very good source for policy analysis.

579. National Research Council. Atmosphere-Biosphere Interactions: Towards a Better Understanding of the Ecological Consequences of Fossil Fuel Combustion. Washington, DC: National Academy Press, 1981.

A relatively bureaucratic report which tries to synthesize published research available in 1980. Of value only because of some of its graphic compilations.

580. ______. Acid Deposition, Atmospheric Processes in Eastern North America, report of the Committee on Atmospheric Transport and Chemical Transformation in Acid Precipitation. Washington, DC: National Academy Press, 1983.

Policy-oriented analysis of scientific reports. Not technical but weakened because of its obvious lack of technical understanding. Very influential in national policy development because of the source.

581. National Research Council of Canada. Acidification in the Canadian Aquatic Environment: Scientific Criteria for Assessing the Effects of Acidic Deposition on Aquatic Ecosystems. Ottawa: NRCC, 1981.

A national policy document of use to policy scientists and analysts. Good for comparison with U.S. documents.

582. Nephew, E.A. "The Challenge and Promise of Coal," Technology Review, Vol. 76, 1973, pp. 21-29.

A relatively non-technical assessment of the potential uses, social benefits, and social and environmental costs of large-scale coal usage in the United States. This is set in an environment of rising oil and gas prices, fuel shortages and a national appeal to switch to coal as our primary fuel. Very useful for policy analysis and as background reading for anyone interested in the changing patterns of industrial coal use in the United States.

583. New England River Basin Commission. The Economic and Social Significance of Acid Deposition in the New England/New York Region. Boston: NERBC, 1981.

A comprehensive examination of the economic impact of acid rain in this most sensitive area. The social consequences are as a result of economic changes brought on by environmental degradation due to acid deposits. A good source of general economic predictions.

584. Newman, L. "Important Considerations on the Incorporation of Sulfur and Nitrogen into Rain," in Proceedings of the Conference on Acid Rain and the Atlantic Salmon. Portland, ME: University of Maine, 1980, pp. 13-15.

A short, abstract-like report on some of the processes and controlling factors involved in the acidification of precipitation. Of general use to all scientists.

585. Nihlgard, B. "Precipitation, Its Chemical Composition and Effect on Soil Water in a Beech and Spruce Forest in South Sweden," Oikos, Vol. 21, 1980, pp. 208-217.

A rather early report on acid precipitation in Swedish forests. The reports of forest effects are at a general rather than specific level but do document general forest health decline with acid precipitation increases. In light of later research, this is a general source of only comparative value.

586. Noble, R.D. and K.F. Jensen. "An Apparatus for Monitoring CO_2 Exchange Rates in Plants During SO_2 and O_3 Fumigation," Journal of Environmental and Experimental Botany, Vol. 34, 1983, pp. 470-475.

A useful technical article describing a technique and equipment configuration that can be used to evaluate chemical exchanges in simulated acid-rain and related combustion-pollutant experiments. A real breakthrough for this type of experiment.

587. Noggle, J.C. "Sulfur Accumulation by Plants: The Role of Gaseous Sulfur in Crop Nutrition," in D.S. Shriner, et al. (eds), Atmospheric Sulfur Deposition. Ann Arbor, MI: Ann Arbor Scientific Publishing, 1980, pp. 289-297.

Excellent, technical chapter on the biochemistry of sulfur ingestion. Implications for acid-rain-induced overdoses are discussed thoroughly.

588. Nordo, J. "Long Range Transport of Air Pollutants in Eruope and Acid Precipitation in Norway," Water, Air, and Soil Pollution, Vol. 6, no. 2, 1976, pp. 199-217.

A European perspective on their international acid problem. Particularly useful for comparison with U.S.-Canada situation. Well written, not overly technical; documents downwind effects of acid rain as far as 1,000 miles away from combustion sources.

589. Norton, S.A. and A. Henriksen. "The Importance of CO_2 in Evaluation of Effects of Acidic Deposition," Vatten, Vol. 39, 1983, pp. 346-354.

A rather technical report on research into the function of CO_2 in the acid rain cycle. Looks at wet and dry deposition on all types of surfaces.

590. Nriagu, J.O. (ed). Sulfur in the Environment, Part II: Ecological Impacts. New York: John Wiley and Sons, 1978.

A very technical analysis of the many ways in which sulfur is introduced into the air, water, and soil. Not specifically oriented toward acid rain but presents much that underlies the acid rain process.

591. Oden, S. The Acidification of Air and Precipitation and Its Consequences in the Natural Environment, Ecology Committee Bulletin No. 1. Stockholm: The State National Science Research Council, 1968.

An early article by one of Europe's leading acid rain authorities. A general overview but with enough technical information to serve as a true reference.

592. Oden, S. and T. Ahl. "The Acidification of Scandinavian Lakes and Rivers," Ymer Arsbok, 1970, pp. 103-122.

A landmark research synthesis for European environmentalists. Can be useful for comparative purposes.

593. Oden, S. "The Acidity Problem--An Outline of Concepts," Water, Air, and Soil Pollution, Vol. 6, 1976, p. 158.

A short report meant to stir research in a scientific audience.

594. _______ and A. Thorsten. "The Sulfur Budget of Sweden," in T.C. Hutchinson and M. Havas (eds), Effects of Acid Precipitation on Terrestrial Ecosystems. New York: Plenum Press, 1980, pp. 111-122.

Traces sulfur from fireplace to forest in Sweden. Rather technical but applicable as a chemical model in any environment.

595. O'Gara, P.J. "Sulfur Dioxides and Fume Problems and Their

Solution," Industrial and Engineering Chemistry, Vol. 14, 1922, p. 744.

A very early look at effluent from coal combustion and some suggestions about dealing with it. Has interesting historical significance.

596. Ohio River Valley Regional Study Group. Air Quality Assessment of the Ohio River Valley Region, report to the National Commission on Air Quality, 1980.

A synthetic baseline study of a primary source region for acid rain. Important set of data.

597. Omernik, J.M. Nonpoint Source-Stream Nutrient Level Relationships: A Nationwide Survey, EPA-600/3-77-105, U.S. Environmental Protection Agency. Corvallis, OR: Environmental Research Laboratory, 1977.

Looks at gross pollution sources for streams and lakes. Primary focus is on land-use patterns, especially agricultural chemicals introduced into watersheds by runoff. Acid rain is given a lesser role.

598. ________ and C.G. Powers. "Total Alkalinity of Surface Waters--A National Map," Annals of the Association of American Geographers, Vol. 47, 1983, pp. 133-136.

A professionally produced map that can serve as a focus of discussion. The data behind it give acid rain a small role in surface water pollution compared to agricultural chemical use.

599. Ontario Ministry of the Environment. A Case Against the Rain. Ottawa: Department of the Environment, 1980.

Government publication meant for general public. Not technical or incisive. Good for international policy analysis.

600. ________. Acid Sensitivity Survey of Lakes, APIOS Report No. AP1 002/81, Toronto, 1981.

Technical report with data sets that can be of use. Little analysis provided but good source of information for comparative study of lake regions.

601. Oppenheimer, M. "The Relationship of Sulfur Emissions to Sulfate in Precipitation," Atmospheric Environment, Vol. 17, 1983, pp. 451-460.

An excellent, thorough examination of the correlation between sulfur emission from point sources with the sulfur compound levels found in precipitation. Sulfur is a major constituent of acid rain and this article provides some very critical insight into the process of atmospheric acidification.

602. Organization for Economic Cooperation and Development. *The Costs and Benefits of Sulphur Oxide Controls--A Methodological Study*. Paris: OECD, 1981.

A comprehensive report of a study of the environmental and socioeconomic effects that may result from stiffer controls of combustion effluent. Because of Europe's heavy dependence on coal, this is particularly important to them and this report focuses on the best ways to carry out a study of these effects. Very useful for sociologists, economists, and public administrators.

603. Ormrod, D.P., et al. "Air Pollution Effects on Agricultural Crops in Ontario: A Review," *Canadian Journal of Plant Science*, Vol. 46, 1980, pp. 1023-1030.

A general assessment of effluent effects on crops in Canada. Most of the analysis is inferential but could be used in a comparative study.

604. Ottar, B. "An Assessment of the OECD Study on Long Range Transport of Air Pollutants," *Atmospheric Environment*, Vol. 12, 1978, pp. 445-454.

A policy-oriented examination of the European attempt to develop an international system of monitoring and evaluating air pollution and its transboundary effects. Good reading as well for scientists who have an interest in national or international networks of air pollution monitoring.

605. Overrein, L.N. "Sulfur Pollution Patterns Observed: Leaching of Calcium in Forest Soil Determined," *Ambio*, Vol. 1, 1972, pp. 145-147.

An early study of acid rain infiltration in a forest environment. Short research report with little real data to share.

606. Pack, D.H. "Precipitation Chemistry Patterns: A Two-network Data Set," *Science*, Vol. 108, 1980, pp. 1143-1145.

A short methodological analysis of the data that are collected by the EPRI and the NADP networks. Very quick look at often conflicting evidence.

607. Page, W.P. and J.M. Gowdy. "The Regional Impacts of Long-Range Air Pollution Transport Associated with PSD Requirements," in D.C. Nichols, et al. (eds). *Energy and the*

Environment: Proceedings of the Sixth National Conference. Dayton, OH: American Institute of Chemical Engineers, 1979, pp. 169-175.

A rather non-scientific examination of potential economic impacts of long-range transport of selected combustion effluents, especially SO_2. The focus is on the region of Indiana, Ohio, Kentucky in the mid-eastern U.S. A general article; no new data or analysis.

608. Painter, D.E. Air Pollution Technology. Reston, VA: Reston Publishing Company, 1977.

Basically a technological description of the machines and systems in use, and being developed, to deal with the causes of air pollution. There is no discussion here of acid rain effects but a comprehensive examination of the technology available to control effluent and thus affect the acid rain cycle at its origin.

609. Parker, L.B. Distributing Acid Rain Mitigation Costs: Analysis of a Three-Mil User Fee on Fossil Fuel Electricity Generation. Washington, DC: U.S. Congressional Research Service, 1983.

An analysis of the economic impact of legislative imposition of a tax on major coal users as a way of generating funds to help solve the problems of acid rain. No new information but some insightful analysis of the national impact such an action would have. Good source for economists and for policy students.

610. _______ and D.A. Thompson. Acid Rain Legislation and the Future of the Eastern Low-Sulfur Coal Industry, CBS Report No. 84-89-ENR. Washington, DC: U.S. Congressional Research Service, 1984.

A general, rather bureaucratic report on the economic and social effects that legislation and regulation aimed at controlling acid rain have and will have on a major coal-producing region in the United States. Of interest primarily to policy students because of the impact CRS has on Congress and thus on federal legislation.

611. Patrick, C.H. "Developmental Toxicology as Input to the Methodology for Human Studies of Energy-Related Pollutants," in D.D. Mahlum, et al. (eds), Developmental Toxicology of Energy-Related Pollutants. Washington, DC: U.S. Department of Energy, 1978, pp. 425-440.

A report on methods used to identify and study prenatal and neonatal diseases believed to be due to exposures to energy-related pollutants. Uses statistics on congenital abnormalities from the Oak Ridge, Tennessee region where two large coal-fired plants, as well as a major nuclear facility, are located. The results show that it is difficult or impossible to define precisely those abnormalities caused by combustion effluent or radiation as opposed to "normally occurring" abnormalities. Good background for students of air pollution and human health.

612. Patrick, R.V., et al. "Acid Lakes from Natural and Anthropogenic Causes," Science, Vol. 211, 1981, pp. 446-448.

A general, almost journalistic article that can serve as background reading. Chronicles all sources of lacustrine acidity.

613. Pearce, F. "The Menace of Acid Rain," New Scientist, Vol. 95, no. 1318, 1982, pp. 419-424.

Journalistic article with editorial qualities. Can be useful to policy studies.

614. Pearse, G.H.K. A National Strategy for By-product Sulphur. Ottawa: Energy, Mines and Resources Canada, 1980.

A comprehensive look at the problems associated with and some solutions offered for the large amounts of sulfur by-products, airborne and terrestrial, that are produced by important Canadian industries.

615. Pearson, F.J. and D.W. Fisher. Chemical Composition of Atmospheric Precipitation in the Northeastern United States, U.S. Geological Survey Water Supply Paper 1535-P. Washington, DC: U.S. Geological Survey, 1971.

Excellent compendium of data at an early date. Used for baseline and comparative study. Technical analysis but data are useful even for non-technicians.

616. PEDCo Environmental, Inc. Analysis of Emission Reductions and Air Quality Changes for Alternative Development Scenarios, report to the National Commission on Air Quality, Contract No. 11u-AQ-7695, November 1980.

Pollution control modeling approach to the problem of sulfur emissions. Technical, not for non-scientists.

617. _______. Analysis of the Emission Reductions, Air Quality Changes and Institutional Impacts Resulting from the Implementation of Selected Policy Alternatives, report to the National Commission on Air Quality, Contract No. 11v-AQ-7695, November, 1980.

Emission modeling study comparing predicted scenarios for different proposed national environmental policies. Technical but not beyond general comprehension. Good for policy analysis.

618. _______. Control Strategies for Coal Fired Utility Boilers, DOE/METC-82-42. Washington, DC: U.S. Department of Energy, 1982.

An examination and analysis of various proposed technological systems designed to reduce effluent from coal combustion. Of interest not only to combustion engineers but to policy scientists because of the implications of this report for federal regulatory action against coal users.

619. Peden, M.E. and L.M. Skowron. "Ionic Stability of Precipitation Samples," Atmospheric Environment, Vol. 12, 1978, pp. 2343-2349.

Technical report documenting chemical dynamics in polluted precipitation at both micro and macro levels of analysis. For scientifically trained readers.

620. Perhac, R.M. "Sulfate Regional Experiment in Northeastern United States: The 'SURE' Program," Atmospheric Environment, Vol. 12, 1978, pp. 641-647.

An introduction to the methodology and utility of an experimental modeling attempt by EPRI. Of use for policy scientists and others.

621. _______, et al. EPRI Sulfate Regional Experiment: Results and Implications. Palo Alto, CA: Electric Power Research Institute, EA-2077-SY-LD, 1981.

Technical report on aerosol sulfur research. The data provided are useful; the analysis is biased by the supporting organization but does indicate alternative explanations that should be considered.

622. Perhac, R.M. Atmospheric Retention of Anthropogenic CO_2: Scenario Dependence on the Airborne Fraction, EA-3466. Palo Alto, CA: Electric Power Research Institute, 1984.

A technical report prepared by a research firm funded by EPRI. Examines rather technically the current models of physical and chemical dynamics of airborne coal combustion effluent.

623. Perry, H. Energy--The Ultimate Resource, Report Serial J of the Task Force on Energy of the Subcommittee on Science, Research, and Development of the Committee on Science and Astronautics, 92nd Congress. Washington, DC: U.S. Government Printing Office, 1971.

An immense compendium of charts, tables, and data arrays on all aspects of the energy production, consumption, and pollution cycle. This predates national concern with acid rain but is a grand source of data on any aspect of energy use in the United States and can be utilized in study of effluent production.

624. Peskin, H.M. "Environmental Policy and the Distribution of Benefits and Costs," in P.R. Portney (ed), Current Issues in U.S. Environmental Policy. Baltimore, MD: Johns Hopkins University Press, 1978, pp. 144-163.

Basically a policy analysis but this focuses on the Clean Air Act and its amendments as they affect federal regulation of the energy-related industries. Provides a non-econometric evaluation of the costs and benefits associated with implementation of the Clean Air Act.

625. Peters, N.E. and J.E. Bonelli. Chemical Composition of Bulk Precipitation in the North-Central and Northeastern United States, December 1980 through February 1981, USGS Circular 874. Washington, DC: U.S. Geological Survey, 1982.

Precipitation data compilation that can serve as the basis for comparative or regional study. Very useful data sets.

626. Peters, N.E., et al. Temporal Trends in the Acidity of Precipitation and Surface Waters of New York, Water Supply Paper no. 2188. Reston, VA: U.S. Geological Survey, 1982.

A thorough examination of data on surface and atmospheric water pH over a rather long period of time as taken from USGS records. No real attempt at analysis but an excellent technical presentation of very usable data.

627. Peterson, I. "Acid Lakes and Thin Soils," Science News, Vol. 123, no. 21, 1983, pp. 332-333.

Short, journalistic report on local geology and lacustrine acidification. Synthetic.

628. Pfeiffer, M.H. and P. Festa. Acidity Status of Lakes in the Adirondack Region of New York in Relation to Fish Resources, FW-P168. Albany, NY: New York State Department of Environmental Conservation, 1980.

Technical study of lacustrine acidity using recognized biological toxicity standards as analysis markers. Good for regional comparisons.

629. Pierson, W., et al. "Ambient Sulfate Measurements on Allegheny Mountain and the Question of Atmospheric Sulfate in Northeastern United States," in T.J. Kneip and P.J. Lioy (eds), Aerosols: Anthropogenic and Natural, Sources and Transport. Albany, NY: New York Academy of Science, 1980, pp. 145-173.

Comprehensive report of field data collection in a forest environment.

Can be a good addition to any compilation of acid deposition case studies.

630. Portney, P.R. (ed). Current Issues in U.S. Environmental Policy, Baltimore, MD: Johns Hopkins University Press, 1978.

A collection from Resources for the Future. A good set of articles on air and water pollution issues, toxic substances, costs and benefits of environmental policy and the issues and implications of regulatory enforcement in the United States. Of real utility for policy analysis.

631. _______. "Acid Rain: Making Sensible Policy," Resources for the Future, No. 75, 1984, pp. 9-12.

An "environmentalist's" perspective on the various economic, social, and political forces active in the development of national policy on acid rain. The specific focus is on employment patterns and the impact acid rain regulation would have on them. Good reading for a general audience.

632. Postel, S. Air Pollution, Acid Rain, and the Future of Forests, Worldwatch Paper 58. Washington, DC: World Watch Institute, 1984.

A concise up-to-date general overview of the present and potential problems acid rain and other combustion-related pollution can create for forest resources. Rather economically oriented but written for a general audience. Very useful graphics and tables.

633. Pough, E.H. "Acid Precipitation and Embryonic Mortality of Spotted Salamanders," Science, Vol. 192, 1976, pp. 57-61.

Non-technical report on the acidic effect on an important ecological indicator species. Good source of comparative study and for use in predicting ecosystemwide effects.

634. Powell, D.C., et al. A Variable Trajectory Model for Regional Assessments of Air Pollution From Sulfur Compounds, Report No. PNL-2734. Richland, WA: Battelle Pacific Northwest Laboratory, 1979.

A very comprehensive, very technical report of tests of statistical/synoptic models of long-range pollution transportation through the atmosphere. Very good for later comparison.

635. Powell, T.G. and R.W. MacQueen. "Precipitation of Sulfide Ores and Organic Matter: Sulfate Reactions at Pine Point, Canada," Science, Vol. 224, no. 4644, 1984, pp. 63-66.

Concise research report on an important regional acid precipitation situation. Very useful information on atmospheric chemistry of smelter effluent.

636. Preston, E.M. National Crop Loss Assessment Network (NCLAN) 1980 Annual Report. Corvallis, OR: U.S. Environmental Protection Agency, 1981.

Details series of research on responses of crop plants to acid rain and dry deposition. Part of an annual cycle of reports that provide excellent comparative study.

637. _______. National Crop Loss Assessment Network (NCLAN) 1981 Annual Report. Corvallis, OR: U.S. Environmental Protection Agency, 1982.

Similar to report above.

638. _______. "The National Crop Loss Assessment Network (NCLAN): An Interdisciplinary Approach to Assessment of the Economic Impact of Air Pollution on Agriculture," Proceedings, 75th Annual meeting of the APCA, 1982, paper no. 82-69.1.

Outline of the administration of NCLAN and the proposed effect this program will have on national policy development and scientific research.

639. Puckett, L.J. "Acid Rain, Air Pollution and Tree Growth in Southeastern New York," Journal of Environmental Quality, Vol. 24, 1984, pp. 34-41.

Good longitudinal study report on forest ecology under the influence of acid rain. Use of tree rings as indicators and regional aspects of the study are of special interest.

640. Radke, L.F., et al. "Precipitation Scavenging of Aerosol Particles," Proceedings of the Conference on Cloud Physics and Atmospheric Electricity. Boston: American Meteorological Society, 1978, paper no. 12.

Very technical paper on rates of chemical interaction in the atmosphere. Basic data to an understanding of the process by which acid rain is formed at a near-molecular level. The information on particle "selection" is very useful for atmospheric modeling.

641. Ragland, K.W. and K.E. Wilkening. "Relationship Between Mesoscale Acid Precipitation and Meteorological Factors," in

F.M. D'Itri (ed), Acid Precipitation: Effects on Ecological Systems. Ann Arbor, MI: Ann Arbor Science Publishers, 1982, pp. 123-140.

A well-written scientific analysis of an atmospheric model for SO_2 deposition and its relationship to upper atmospheric mixing. Based on data from the Rainy Lake region of Minnesota. Seasonality taken into account. Good regional isopleth maps of all variables that are useful for inference to other regions.

642. Rahn, K.A. and R.J. McCaffrey. "On the Origin and Transport of the Winter Arctic Aerosol," Annals of the New York Academy of Science, Vol. 338, 1980, pp. 486-503.

Rather technical but well-written article on transportation of atmospheric pollution into the Arctic region. Transport models presented here can be applied to other polar regions.

643. Rahn, K.A., et al. "High Winter Concentration of SO_2 in the Norwegian Arctic and Transport From Eurasia," Nature, Vol. 287, 1980, pp. 824-826.

Research report on atmospheric monitoring program in an arctic region. Data on acidic and pre-acidic atmospheric chemicals are very useful; information and speculation on the transport distances and mechanisms for this pristine region are also important.

644. Rampino, M.R. and S. Self. "The Atmospheric Effects of El Chichon," Scientific American, Vol. 250, no. 1, 1984, pp. 133-140.

An in-depth report of analysis of the atmospheric pollution created by the recent volcanic eruption. The cloud created was as thick as that at Krakatoa and the results of this analysis can be compared well to what is known and speculated about the atmospheric impact of all volcanic activity.

645. Rasmussen, R.A. and M.A.K. Khalil. "Carbon Monoxide in the Earth's Atmosphere: Increasing Trend," Science, Vol. 224, no. 4644, 1984, pp. 54-56.

Results of a multi-year study and data collection project carried out in Oregon. Data are very useful although rather regionally specific.

646. Raynal, D.J., et al. Actual and Potential Effects of Acid Precipitation on Forest Ecosystems in the Adirondack Mountains, Report 80-28. Albany, NY: Environmental Research and Development Authority, 1980.

Well-documented technical report and predictive analysis of forest effects in one of the most sensitive pristine regions in the U.S. Good source for scientists and non-scientists alike.

647. _______. Characterization of Atmospheric Deposition and Ionic Input at Huntington Forest, Adirondack Mountains, New York, ESF 83-003. Albany, NY: Institute of Environmental Program Affairs, 1983.

A complete, in-depth report of continuing research on acidic deposition in the Adirondacks. This report is at a more refined level of analysis than is previous work from this author due to continued refinement of the collection and analysis process with time. Very useful data, regionally specific analysis.

648. Raynor, G.S., et al. "Particulate Dispersion Into and Within a Forest," Boundary-Layer Meteorology, Vol. 7, 1974, pp. 429-456.

A very technical report on the movement of pollutant material into and through a forest environment. Provides excellent processional models; very little data of direct application to acid rain study. Good source for forest ecologists.

649. Raynor, G.S. "Meteorological and Chemical Relationships From Sequential Precipitation Samples," in W. Licht, et al., (eds), American Institute of Chemical Engineering Symposium Series No. 188, Vol. 75. New York: AIChE, 1979, pp. 269-273.

Very technical report on physical and chemical dynamics that underlay the production and deposition of acidic precipitation.

650. Rebbeck, J. and E. Breenan. "The Effect of Simulated Acid Rain and Ozone on the Yield and Quality of Glasshouse-Grown Alfalfa," Environmental Pollution, Vol. 36, no. 1, 1984, pp. 7-16.

Report on tests using different levels of acid in applications. Some testing of diazanon as a depositional retardant also. Overall results indicate no significant effect on either yield or quality of alfalfa. A good set of data that substantiates the idea that there is very little direct effect of acid rain on most plants. Good source for botanists and agronomists.

651. Regens, J.L. "Acid Rain: Does Science Dictate Policy or Policy Dictate Science?" in T.D. Crocker (ed), Economic Perspectives on Acid Deposition Control. Boston: Butterworth Publishers, 1984, pp. 5-20.

An interesting, almost philosophical analysis of the way in which science and policy have affected each other on the acid rain issue. Non-technical but scholarly.

652. Reich, P.B., et al. "Reduction in Soybean Yield After Exposure to Ozone and Sulfur Dioxide Using a Linear Gradient

Exposure Technique," Water, Air, and Soil Pollution, Vol. 17, 1982, pp. 29-36.

Research report on combustion effluent effects on one of the most important agricultural crops grown in the U.S. Good baseline information on general results. No specific physiological information provided.

653. Reinert, R.A., et al. "Plant Responses to Pollutant Combinations," in J.B. Mudd and T.T. Kozlowski (eds), Responses of Plants to Air Pollution. New York: Academic Press, 1975, pp. 212-221.

A technical article on simulated tests of plants with mixtures of possible air pollutants applied both directly and indirectly. Data on plant responses are very well presented and useful for comparative study.

654. Renberg, I. and T. Hellberg. "The pH History of Lakes in Southwestern Sweden, as Calculated From the Subfossil Diatom Flora of the Sediments," Ambio, Vol. 11, no. 1, 1982, pp. 30-32.

A short technical report on study of paleo-sediments as a method of determining micro-life populations in the past and inferring changes in lacustrine pH from those population changes. This report makes some inferences that are highly theoretical but which, as more data on related phenomena are gathered, are proving to be very accurate. Good source for ecologists and geologists.

655. Rentschler, I. "Significance of the Wax Structure in Leaves for the Sensitivity of Plants to Air Pollutants," Proceedings of the Third International Clean Air Congress. Dusseldorf, GDR: OECD, 1973, pp. A139-A142.

A rather scientific report on the ways in which plant physiology, specifically the thickness and resistance of cuticular waxes on leaves, can affect a plant's susceptibility to aerosol pollutants. A precursor to later work by Shriner and others.

656. Retzsch, W.C., et al. Alternative Explanations for Aquatic Ecosystems Effects Attributed to Acidic Precipitation, consultant report by Everett and Associates, Rockville, Maryland, March, 1982.

Analytic report supported by the electric utilities to focus attention away from acid rain as a major resource of pollution in water resources. Raises good points for discussion of the actual sources of acid precipitation.

657. Reuss, J.O. Chemical/Biological Relationships Relevant to

Ecological Effects of Acid Rainfall. Corvallis, OR: National Environmental Research Center, 1975.

A general treatise on the overall environmental effects of acid rain. Early study, dependent on general rather than specific research but many of the relationships predicted or inferred have been found to be real.

658. Rhodes, S.L. and P. Middleton. "The Complex Challenge of Controlling Acid Rain," Environment, Vol. 25, no. 4, 1983, pp. 31-38.

Policy-oriented article for general audience. Presents economic, political, and scientific considerations for discussion. Offers different scenarios for regulation approaches with hypothesized results. Very useful for economists and policy students.

659. Rhodes, S.L. "Superfunding Acid Rain Controls: Who Will Bear the Costs?" Environment, Vol. 26, no. 6, 1984, pp. 25-32.

A policy-oriented article focusing on the possible use of the EPA "superfund" of industrial polluter fines to pay for control of acid rain. A central question raised is how to isolate the specific polluter when acid rain data are regional at best. Very useful for policy students and thought provoking for all environmental scientists.

660. Rice, H., et al. "Contribution of Anthropogenic and Natural Sources to Atmospheric Sulfur in Parts of the United States," Atmospheric Environment, Vol. 15, 1981, pp. 1-9.

A very useful overview of the processes that integrate to form acidic precipitation. This does provide a basic delineation between manmade and natural pollution sources.

661. Rice, R.E. The Effects on Forest and Crop Resources in the Eastern United States. Washington, DC: The Wilderness Society, 1983.

A thorough analysis of published research with a focus on the overall economic effects of acid rain on the harvested plants of the Eastern U.S. No new data but it does serve to translate data from many reports into a cohesive comparable whole. Good source for policy students.

662. Richter, D.D., et al. "Chemical Composition and Spatial Variation of Bulk Precipitation at a Coastal Plain Watershed in South Carolina," Water Resources Research, Vol. 19, no. 1, 1983, pp. 34-43.

A thorough case study report on a small region in the southern U.S.

Excellent source of data for regional comparison and a good analysis of the geographic parameters of precipitation acidity. Quite useful for climatologists and for general ecologists with a regional approach.

663. ______. "Atmospheric Sulfur Deposition, Neutralization and Ion Leaching in Two Deciduous Forest Ecosystems," Journal of Environmental Quality, Vol. 12, no. 2, 1983, pp. 119-125.

A good scientific report on field research on soil chemistry in deciduous forests under acidic precipitation conditions. The data from the two forests provide internal comparison and the whole article can be used comparatively as well. Good source of data sets for forest ecologists.

664. Ridder, T.B. and A.J. Frantzen. "Acid Precipitation Over the Netherlands," in S. Beilke and A.J. Elshout (eds), Acid Deposition. Boston: Reidel Publishing, 1983, pp. 123-141.

Primarily an examination of the Dutch precipitation chemistry sampling network. Some trend data in acidity and ionic composition are provided as are very usable graphics of these data. Most analysis focuses on the network rather than the data, trends, or atmospheric chemistry.

665. Ridker, R.G. Economic Costs of Air Pollution. New York: Praeger Press, 1967.

A rather econometric analysis of the overall costs of air pollution. Includes health, ecology, and prevention costs. Does not focus specifically on acid rain or related phenomena but does include this as one area of impact. The early publication data preclude any really useful inclusion of acid rain impact data.

666. Riekerk, H., et al. "Atmospheric Deposition Patterns in North-Central Florida," in A.E.S. Green and W.H. Smith (eds), Acid Deposition: Causes and Effects. Gainesville, FL: University of Florida Press, 1983, pp. 90-96.

A short report on micro-regional deposition chemistry in the southern U.S. Really a preliminary report but very useful presentation of data and some analysis. The water resources of this region make it particularly susceptible to acid effects.

667. Rind, D. "Global Climate in the 21st Century," Ambio, Vol. 13, no. 3, 1984, pp. 148-151.

A concise set of global climatological projections, especially focusing on CO_2 and other anthropogenic sources of climatic change and modification. Very well written and useful for a wide audience although the allusion to a U.S.-Canadian "dust bowl" may be a bit overstated for dramatic purposes.

668. Rippon, J.E. "Studies of Acid Rain on Soils and Catchments," in T.C. Hutchinson and M. Havas (eds), Effects of Acid Precipitation on Terrestrial Ecosystems. New York: Plenum Press, 1978, pp. 499-524.

An excellent report of early research into the effects of acid rain on the soil and rocks of certain regions. The author makes clear implications for variable regional effects from acid rain due to susceptible and resistant local and regional geology. Particularly useful for the good data on soils.

669. Rist, D.L. and J.W. Lorbeer. "Ozone-Enhanced Leaching of Onion Leaves in Relation to Lesion Production," Phytopathology, Vol. 74, no. 10, 1984, pp. 1217-1220.

A scientific research report documenting foliar damage to an important vegetable plant under simulated contamination conditions. Ozone-related effects are closely related to other combustion effluent problems, including acidic precipitation.

670. Roberts, L. "California's Fog Is Far More Polluted Than Acid Rain," BioScience, Vol. 32, no. 10, 1982, pp. 778-779.

Uses acid rain more for comparative purposes than for explanation. Does open discussion of deposition through fog which may be a localized California phenomenon.

671. Robinson, E. and R.C. Robbins. "Gaseous Sulfur Pollutants From Urban and Natural Sources," Journal of the Air Pollution Control Association, Vol. 20, 1970, pp. 233-235.

A much too short presentation of the various natural and manmade pollution sources which produce acid rain. For a general audience but lacking much usable data or analysis for a scientist.

672. Robinson, G.D., et al. "Differences in Low pH Tolerance Among Strains of Brook Trout," Journal of Fishery Biology, Vol. 8, 1976, pp. 5-17.

Good, scientific technical report documenting significant differences in response of fish to acidification. Ecological implications are provided.

673. Rodhe, H. "A Study of the Sulfur Budget for the Atmosphere Over Northern Europe," Tellus, Vol. 24, 1972, pp. 128-138.

A very early report on research into the atmospheric chemistry of acid-producing products in Europe. The focus is on the levels of sulfur and the transport of sulfur through the atmosphere rather than on acid rain but the data presented, as well as the analysis of the atmospheric processes, are basic to an understanding of the acid rain phenomenon.

674. _______. "Budgets and Turnover Times of Atmospheric Sulfur Compounds," Atmospheric Environment, Vol. 12, 1978, pp. 671-680.

A very technical analysis of the atmospheric chemistry of various combustion effluent products. Provides some basic data and analysis for further development of transport and deposition models of acid-creating chemicals. Not effects-oriented but this work can be used to infer effects.

675. Roemer, F.G., et al. "Preliminary Measurements from an Aircraft into the Chemical Composition of Clouds," in S. Beilke and A.J. Elshout (eds), Acid Deposition. Boston: Reidel Publishing, 1983, pp. 195-203.

Presents a report on an experiment in the Netherlands with cloud chemistry sampling from aircraft. The methodological analysis is useful as are the data presented from the tests. Particularly interesting comparisons of cloud chemistry with that of resultant precipitation; they do not always relate well.

676. Rohbock, E. "Atmospheric Removal of Airborne Metals by Wet and Dry Deposition," in H.W. Georgii and J. Pankrath (eds), Deposition of Atmospheric Pollutants. Boston: Reidel Publishing, 1982, pp. 159-172.

Report on atmospheric wash-out of heavy metals from polluted air at thirteen sites in Germany. Wet deposition is found to be more significant in the wash-out of small particles; dry deposition found to be of equal importance for larger particles. Because of the close relationship between acid rain processes and heavy metal deposition, these data and analysis are very useful.

677. Roman, J.R. and D.J. Raynal. "Effects of Acid Precipitation on Vegetation," in Actual and Potential Effects of Acid Precipitation on a Forest Ecosystem in the Adirondack Mountains, Report ERDA 80-28. Albany, NY: New York State Energy Research and Development Authority, 1981, pp. 4-1--4-63.

Comprehensive research based on data gathered over several years in upstate New York forest environments. Very useful data on precipitation, water resources, and forest ecology.

678. Ronneau, C. and J.P. Hallet. "Heavy Elements in Acid Rain," in S. Beilke and A.J. Elshout (eds), Acid Deposition. Boston: Reidel Publishing, 1983, pp. 149-154.

Examines the levels of heavy metal pollution contained in and indicated by acid rain in European situations. Traces these metals through the ecosystem into cows' milk in Belgium. Very good preliminary examination of a process related to acid rain but with perhaps more long-term consequences.

679. Rosencranz, A. and G. Wetstone. "Acid Precipitation: National and International Responses," Environment, Vol. 22, no. 5, 1980, pp. 40-43.

A policy-oriented discussion of national and international concerns about acid rain. Legal, regulatory, and policy analysis provided.

680. Rosencranz, A. "Acid Precipitation: The Problem of Transboundary Pollution," Environment, Vol. 22, no. 5, 1980, pp. 15-20.

A general overview of legal, political, and economic problems created by U.S.-Canadian acid rain phenomena. An excellent source for national policy students.

681. Rosenquist, I.T. "Alternative Sources for Acidification of River Water in Norway," The Science of the Total Environment, Vol. 10, 1978, pp. 39-49.

Further exploration of the ways in which acids reach streams. A Norwegian example of land-use and agricultural runoff as co-culprit with acid rain.

682. ________. "Acid Precipitation and Other Possible Sources for Acidification of Rivers and Lakes," The Science of the Total Environment, Vol. 10, 1978, pp. 271-272.

More delineation of the part acid rain plays in stream pollution versus that for agricultural chemicals. A useful discussion that points to areas of needed research in this very important discipline.

683. ________. "Alternative Sources for Acidification of River Water in Norway," Science of the Total Environment, Vol. 10, 1979, pp. 39-44.

Very similar to the reference cited above. Fewer data provided here but analysis and data presentation is perhaps more usable.

684. Rubec, C.D.A. Characteristics of Terrestrial Ecosystems Impinged by Acid Precipitation Across Canada, Working Paper 19. Ottawa: Environment Canada, 1981.

General, comprehensive overview of acid rain effects in Canada. Based on scientific studies but written for a non-scientific audience. Good material for international comparisons.

685. Rubin, E.S. "International Pollution Control Costs of Coal-Fired Power Plants," Environmental Science and Technology, Vol. 17, no. 8, 1983, pp. 366A-377A.

An econometric look at costs and benefits involved with improved pollution control of electric power plants. Good for policy analysis and for study of regulation development and implementation.

686. Sage, B. "Acid Drops From Fossil Fuels," New Scientist, Vol. 85, no. 1197, 1983, pp. 743-745.

A short, almost journalistic report on the relationship between coal chemistry and the resultant acid deposition from coal combustion. Not data oriented.

687. Samson, P.A. "Trajectory Analysis of Summertime Sulfate Concentrations in the Northeastern United States," Journal of Applied Meteorology, Vol. 17, 1980, pp. 1375-1389.

Transport model for movement of effluent and resultant acids across the most polluted region of the U.S. Statistical, inferential with no real field evidence but useful nonetheless to atmospheric modeling.

688. Samson, P.J. On the Linearity of Sulfur Dioxide to Sulfate Conversion in Regional Scale Models, report to the U.S. Office of Technology Assessment. Washington, DC: Government Printing Office, 1982.

Technical analysis of chemical models developed to predict acid rain levels. Has scientific portions but is usable by policy analysts as well.

689. Scheider, W.A., et al. Reclamation of Acidified Lakes Near Sudbury, Ontario. Ottawa: Ontario Ministry of the Environment, 1975.

A comprehensive report on attempts to de-acidify several lakes near the Sudbury smelter complex. The results are mixed but very informative; this report can be very useful as experimental data for any planning to reverse acidification effects.

690. Schindler, D.W. "Experimental Acidification of a Whole Lake: A Test of the Oligotrophication Hypothesis," in D. Drablos and A. Tollan (eds), Proceedings of the International Conference on the Ecological Impact of Acid Precipitation. Oslo: SNSF Project, 1980, pp. 370-374.

A concise report on attempts to de-acidify an entire lake artificially. The results are quite mixed but of some use to students of environmental reclamation or recovery. Similar to the report by the same author below.

691. _______, et al. "Experimental Acidification of Lake 223, Experimental Lakes Area: Background Data and the First Three Years of Acidification," Canadian Journal of Fisheries and Aquatic Science, Vol. 37, 1980, pp. 88-96.

Report on research in an idealized lacustrine situation. Very technical and important for limnological analysis; a "must read" for anyone interested in ecological reclamation. Similar data to report above by the same author.

692. Schlesinger, W.H. and M.M. Hasey. "The Nutrient Content of Precipitation, Dry Fallout, and Intercepted Aerosols in the Chaparral of Southern California," American Midland Naturalist, Vol. 103, no. 1, 1980, pp. 114-122.

Looks at acid rain as a possible positive effect in a desert environment. Good material for comparative study with a wet region forest.

693. Schmitt, G. "Seasonal and Regional Distribution of Polycyclic Aromatic Hydrocarbons in Precipitation in the Rhein-Main-area," in H.W. Georgii and J. Pankrath (eds), Deposition of Atmospheric Pollutants. Boston: Reidel Publishing, 1982, pp. 133-142.

Technical report of the atmospheric process of transport and deposit of another set of combustion-produced pollutants. Seasonal variations are significant in this German example. Not acid rain but the information on transportation is closely related.

694. Schnoor, J.L. (ed). Modeling of Total Acid Precipitation Impacts. Boston: Butterworth Publishers, 1984.

Compilation of very technical chapters on economic and scientific effects of acid rain. Excellent for technical or scientific readers.

695. Schofield, C.L. "Effects of Acid Precipitation on Fish," Ambio, Vol. 5, 1976, pp. 228-230.

General article bringing together results of other research on this topic. Best used as a historical marker. Some very useful data sets on fish population and speciation provided.

696. Schofield, C.L. "Processes Limiting Fish Populations in Acidified Lakes," in D.S. Shriner, et al. (eds), Atmospheric Sulfur Deposition. Ann Arbor, MI: Ann Arbor Scientific Publishers, 1980, pp. 345-356.

Very scientific analysis of lacustrine ecology, based on author's own research and that of others. A landmark chapter for students of fishery science. Well-documented correlations between various aspects of lacustrine chemistry and the reactions of specific fish populations.

697. _______ and J.R. Trojnar. "Aluminum Toxicity to Brook Trout in Acidified Waters," in T.Y. Toribara, et al. (eds), Polluted Rain. New York: Plenum Press, 1980, pp. 341-362.

Aluminum toxicity is related to acidity; good scientific analysis of the relationship is given here. Rather scientific.

698. Scholle, S. "Acid Deposition and the Materials Damage Question," Environment, Vol. 25, no. 8, 1983, pp. 25-32.

A general article outlining what is known of acidic precipitation (both wet and dry) effects of building materials. Also some speculation about potential damage to buildings over time given several pollution-level scenarios.

699. Scott, B.C. "Parameterization of Sulfate Removal by Precipitation," Journal of Applied Meteorology, Vol. 17, 1978, pp. 1375-1389.

Rather technical and specialized report of one segment of atmospheric chemistry which underlies the production of acid rain. Useful for physicists, chemists, and geographers who need to understand the acid rain process.

700. _______. "Sulfate Washout Ratios in Winter Storms," Applied Meteorology, Vol. 20, 1981, pp. 619-625.

Very technical report of scavenging of pre-acidic materials in winter storms. Good source of data and of models for application elsewhere. Seasonal variation in both acid rain amount and in the acid making process discussed.

701. _______. "Predictions of In-cloud Conversion Rates of SO_2 to SO_4 Based Upon a Simple Chemical and Kinematic Storm Model," Atmospheric Environment, Vol. 16, 1982, pp. 1735-1752.

A clear presentation of a model designed to allow prediction of atmospheric acidity. A good basis for further research and analysis or for application in atmospheric physics.

702. Scott, W.D. and P.V. Hobbs. "The Formation of Sulfate in Water Droplets," Journal of the Atmospheric Sciences, Vol. 24, 1967, pp. 54-57.

Deals with atmospheric physics involved with coalescence of aerosol chemicals and water vapor. Short but technical article basic to understanding formation of acid rain.

703. Scott, W.D. "Pollutant Scavenging in Winter Storms," MAP3S Progress Report for FY 1977 and FY 1978. Washington, DC: U.S. Department of Energy, 1979, pp. 257-263.

Meteorological study of the way in which mid-latitude cyclones can take in and transport combustion effluent over large geographical areas. Rather technical but of importance to scientists.

704. Sehmel, G.A. "Particle and Gas Dry Deposition," Atmospheric Environment, Vol. 14, 1980, pp. 983-1011.

A very technical analysis of the chemistry and physics of atmospheric deposition of various pollutants. Not specific to acid rain but concerned with the general processes which underlie the creation and deposition of acid compounds.

705. Seigneur, C., et al. "Formation and Evolution of Sulfate and Nitrate Aerosols in Plumes," in M.M. Benarie (ed), Atmospheric Pollution 1982. New York: Elsevier Publishing Co., 1982, pp. 283-292.

Presents a mathematical model that describes the behavior of different aerosol chemicals in coal smokestack plumes. The author deals with the creation of effluent, its transport and deposition in three different regional environments. Good analysis for atmospheric chemists and physicists.

706. Seip, H.M. and A. Tollan. "Acid Precipitation and Other Possible Sources for Acidification of Rivers and Lakes," The Science of Total Environment, Vol. 10, 1978, pp. 253-270.

A general examination of land-use patterns, groundwater flow, and basin runoff as alternative causes of stream and lake acidification. No new material but a fair overview of this phenomenon.

707. Seip, H.M. "Acid Snow--Snowpack Chemistry and Snowmelt," in T.C. Hutchinson and M. Havas (eds), Effects of Acid Precipitation on Terrestrial Ecosystems. New York: Plenum Press, 1980, pp. 77-94.

Technical chapter on the physics and chemistry of incorporation, storage, and release of acid by snow and snowpack. Good scientific article on what can be a long-term problem.

708. Sequeria, R. "Acid Rain: Some Preliminary Results From Global Data Analysis," Geophysical Research Letters, Vol. 8, 1981, pp. 147-150.

No new information presented here but the global compilation is almost unique. Good source of general information and analysis.

709. ________. "Acid Rain: An Assessment Based on Acid-Base Considerations," Journal of Air Pollution Control Association, Vol. 32, 1982, pp. 241-245.

An applied look at the chemistry underlying acid rain and its control. Technical but not overly so.

710. Sevaldrud, I.H., et al. "Loss of Fish Populations in Southern Norway: Dynamics and Magnitude of the Problem," Proceedings of the International Conference on Ecological Impact of Acid Precipitation. Sandefjord, Norway: SNSF, 1980, pp. 350-351.

A short, almost abstract, report on fish population declines in lakes affected by acid deposition. The data confirm what other studies have shown: loss of species and general decline of fish numbers. A useful set of data for regional comparisons.

711. Shannon, J.D. "A Model of Regional Long-term Average Sulfur Atmospheric Pollution, Surface Removal, and Net Horizontal Flux," Atmospheric Environment, Vol. 13, 1981, pp. 1155-1163.

An interesting presentation of one model for predicting atmospheric movement of effluent. Rather technical and specialized but good for meteorological discussion.

712. Sharpe, W.E., et al. "In Situ Bioassays of Fish Mortality in Two Pennsylvania Streams Acidified by Atmospheric Deposition," Northeastern Environmental Scientist, Vol. 2, no. 324, 1983, pp. 171-178.

Although the title implies it, there is no positive connection made between atmospheric deposition and stream acidity in this report. The study on fish mortality is quite complete and negative regarding the impact of acidity in streams. Very useful data and interspecific comparisons. Rather scientific.

713. Shaw, R.W. "Acid Precipitation in Atlantic Canada," Environmental Science and Technology, Vol. 13, no. 4, 1979, pp. 406-411.

A general article outlining the known and potential environmental effects of acid rain in the maritime provinces as of 1979. More speculation than would be warranted now.

714. Shepherd, A.A. and D.H. Pack (eds). Proceedings: Advisory Workshop on Methods for Comparing Precipitation Chemistry Data. Richland, WA: Sigma Research, 1983.

Collection of papers presented at a workshop on precipitation monitoring methodology. Funded by the electric utilities and written with an obvious bias. Useful for discussion.

715. Shinn, J.H. and L. Scott. "Do Man-Made Sources Affect the Sulfur Cycle of Northeastern States," Environmental Science and Technology, Vol. 13, no. 9, 1979, pp. 1062-1067.

An analysis of anthropogenic influence on the movement of sulfur through the atmosphere and the hydrologic cycle. The answer is yes. Good overview article; not too technical.

716. Shriner, D.S., et al. "Simulated Acidic Precipitation Causes Direct Injury to Vegetation," Proceedings of the American Phytopathology Society, Vol. 1, 1974, p. 112.

A short research note on the effects of acid precipitation on cuticular waxes and leaf surfaces in general. Good footnote.

717. Shriner, D.S. "Effects of Simulated Rain Acidified With Sulfuric Acid on Host-Parasite Infections," Water, Air, and Soil Pollution, Vol. 8, 1977, pp. 133-143.

A definitive article for agricultural scientists. Laboratory analysis of acid effects on nitrogen-fixing bacteria associated with soybeans has become the background for much subsequent analysis. Very good for scientist and non-scientist alike.

718. _______. "Effects of Simulated Acidic Rain on Host-Parasite Interactions in Plant Diseases," Phytopathology, Vol. 68, 1978, pp. 213-218.

A similar report to others mentioned above. The focus this time is harmful parasites only; primarily those found on leaf surfaces or which spend some of their life cycle on a leaf surface and then become susceptible to acid rain directly.

719. _______, et al. (eds). Atmospheric Sulfur Deposition: Environmental Impact and Health Effects. Ann Arbor, MI: Ann Arbor Science Publishers, 1980.

A grand collection of chapters by leading scientists in this subject. The field of emphasis is a little too broad for the book to hold up as a coherent unit but each chapter is strong enough to stand on its own. The best single book on acid rain now available.

720. Shriner, D.S. "Vegetation Surfaces: A Platform for Pollutant/Parasite Interactions," in T.Y. Toribara, et al. (eds), Polluted Rain. New York: Plenum Press, 1980, pp. 259-272.

A thorough examination of the ways in which deposited pollutants can affect the ways in which airborne biological materials, primarily fungi and bacteria, affect plants. Also looks at the direct chemical effects of polluted rain or other aerosols deposited on leaf surface pathogens. No new research but an excellent analysis of these processes utilizing previous research.

721. _______ and J.W. Johnston. "Effects of Simulated Acidified Rain on Nodulation of Leguminous Plants by Rhizobium Species," Environmental and Experimental Botany, Vol. 21, 1981, pp. 199-209.

An in-depth presentation of Shriner's dissertation research on soybeans. Good information for soil scientists, agronomists, and biologists.

722. Siccamma, T.G. and W.H. Smith. "Lead Accumulation in a Northern Hardwood Forest," Environmental Science and Technology, Vol. 12, 1978, pp. 593-594.

A relatively non-technical examination of secondary chemical pollution caused by acidified precipitation percolating through forest soils.

723. Siccama, T.G., et al. "Decline of Red Spruce in the Green Mountains of Vermont," Bulletin of the Torrey Botanical Club, Vol. 109, 1982, pp. 162-168.

Clearly written research report documenting forest decline in New England. Acid rain is one of the major reasons cited although research reported on is just on the rate of decline, not on mechanisms behind that decline. Written for an informed general public.

724. Singh, J.J. and A. Deepak (eds). Environmental and Climatic Impact of Coal Utilization. New York: Academic Press, 1980.

A comprehensive collection of chapters concerned with a wide variety of environmental problems created by increased coal use. Acid rain is the primary focus of the climatic impact chapters and some excellent scientific analyses are forthcoming. The non-atmospheric problems associated with coal mining and use are also examined. A well-rounded reference for environmental scientists.

725. Singh, B.R. et al. "Effect of Simulated Acid Rain on Sulfate Movement in Acid Forest Soils," Soil Science Society of America Journal, Vol. 44, 1980, pp. 75-80.

Comprehensive research report on the ionic dynamics within a forest soil under the effects of acid rain. Particularly important because of the relative dearth of knowledge in this area of science. Rather technical.

726. Sisterson, D. and B. Worfel. Seasonal and Annual Comparison of Weekly and Event Sampling of Precipitation Chemistry, Report ANL-81-85. Chicago: Argonne National Laboratory, 1982.

Rather technical assessment of methods used to collect and analyze precipitation samples for chemical makeup. Very good source of data collected in several programs.

727. Skarby, L. and G. Sellden. "The Effects of Ozone on Crops and Forests," Ambio, Vol. 13, no. 3, 1984, pp. 68-72.

A very useful article with an effects orientation. While ozone holds the title, CO_2, SO_2 and NO_2 are all discussed with respect to their impact on important plants. As the authors relate, 90% of crop and forest damage in the U.S. are a result of these combustion created chemicals and this examination is, if for no other reason, worth reading.

728. Skowron, M. and M.E. Peden. "Ionic Stability of Precipitation Samples," Atmospheric Environment, Vol. 12, 1978, pp. 2343-2349.

Examines the basic physical chemistry of standing water collected for chemical analysis. Very useful for determining precipitation collection methods and techniques.

729. Slanina, J., et al. "Investigation of the Source Regions for Acid Deposition in the Netherlands," in S. Beilke and A.J. Elshout (eds), Acid Deposition. Boston: Reidel Publishing, 1983, pp. 129-141.

Preliminary results from a study of a single power plant in the Netherlands. Excellent data of plume washout and of depositional patterns from a major pollution source. The results indicate a more regional set of effects than is often predicted in North America.

730. Slinn, W.G.N. "Estimates of Wet, Dry and Total Deposition of Atmospheric Particles and Gases to a Forested Canopy," Proceedings of the First International Symposium on Acid Precipitation and the Forest Ecosystem held in Columbus, Ohio, 1975. Columbus, OH: Ohio State University, 1976, pp. 113-119.

Analysis based on ecological models rather than on any specific field research. Interesting most as a starting point for future field study.

731. _______. "Precipitation Scavenging," in D. Randerson (ed),

Atmospheric Sciences and Power Production. Washington, DC: U.S. Department of Energy, 1983, pp. 45-51.

A well-written, well-documented report of the way in which precipitation takes material from the atmosphere. Important to an understanding of the production of acid rain.

732. Smiley, R.W. and R.J. Cook. "Use and Abuse of the Soil pH Measurement," Phytopathology, Vol. 62, 1972, pp. 193-194.

A methodological note useful for soil scientists only. It does bring to the surface most of the soil-related methodological problems however and is of great utility to anyone planning a field experiment in this area.

733. Smith, F.B. and R.D. Hunt. "Meteorological Aspects of the Transport of Pollution Over Long Distances," Atmospheric Environment, Vol. 12, 1978, pp. 461-477.

Presents a basic understanding of the physics and chemistry of transport of effluent through the atmosphere. Rather technical but gives a concise overview of the beginning of the acid rain process.

734. Smith, R.J. "Acid Rain Agreement," Science, Vol. 209, 1980, p. 890.

A short policy note on the guidelines of an agreement between the U.S. and Canada. Of use in policy history.

735. Smith, T.B., et al. "Transport of SO_2 in Power Plant Plumes: Day and Night," Atmospheric Environment, Vol. 12, 1978, pp. 605-611.

Excellent report of research into the dynamics of the movement of sulfur through the atmosphere into the acid rain cycle. Very useful information for atmospheric scientists.

736. Smith, W.H. "Air Pollution--Effects on the Structure and Function of Plant-surface Microbial-ecosystems," in T.O. Johnson (ed), Microbiology of Aerial Plant Surfaces. New York: Academic Press, 1976, pp. 75-105.

A rather scientific examination of the ways in which the microbiological communities on leaf surfaces react to chemicals deposited from polluted skies. Acid rain included but is not the only pollutant utilized. Good reference for phytopathologists.

737. ________. Air Pollutants and Forests: Interactions Between Air Contaminants and Forest Ecosystems. New York: Springer-Verlag, 1981.

A very useful book on a wide range of sub-topics focused generally on forests reacting to pollutants. The author relies heavily on personal research experience but manages to bring in a comprehensive set of reports from others to provide a true overview of the processes and problems associated with pollution in forests. Acid rain and deposition are covered but are not the central theme of the book. Not overly technical; useful for a wide audience of informed readers.

738. Sochasky, L. (ed). Acid Rain and the Atlantic Salmon. St. Andrews, New Brunswick: International Atlantic Salmon Foundation, 1981.

A compilation of scientific, economic and political essays on the effects of acid rain on a major Canadian fishery. A good source of specialized information for comparative analysis.

739. Stensland, G.J. "Acid Precipitation in Central Illinois," MAP3S Progress Report for FY 1977 and FY 1978. Washington, DC: U.S. Department of Energy, 1979, pp. 135-140.

Data sets from early precipitation monitoring sites. Not as comprehensive as NADP data but can be spliced into that continual record.

740. ________. NADP Quality Assurance Report, 1 January 1979 to 31 December 1979, Central Analytical Laboratory Report, 1980.

Early NADP data sets along with corrections of preliminary data released earlier. Also includes methodological notes of interest.

741. ________. "Precipitation Chemistry Trends in the Northeastern United States," in T.Y. Toribara, et al. (eds), Polluted Rain. New York: Plenum Press, 1980, pp. 87-104.

A good regional compilation of information from original and secondary sources. Looks at methods of calculating precipitation pH with an eye toward a national norm. Also examines an Illinois site as a case study for changes in precipitation pH over time. Finally, he looks at the Illinois site and others to the East for relationships between sulfate and nitrate deposits and upwind emission trends. Good general article.

742. ________ and R.G. Semonin. "Another Interpretation of the pH Trend in the United States," Bulletin of the American Meteorological Society, Vol. 63, 1982, pp. 1277-1284.

Further examination of the data on environmental acidification with an eye toward comparison with natural cycles of acidification.

743. Stern. A.C. "Air Pollution Control at the Next Technological

Plateau," in K.E. Noll, et al. (eds), Air Pollution Control and Industrial Energy Production. Ann Arbor, MI: Ann Arbor Science Publishers, 1975, pp. 1-4.

A very short, almost introductory chapter on the economics and technological parameters of future air pollution policy development. Of use primarily as an early benchmark in air pollution policy thinking.

744. Stimaitis, D., et al. On-Site Meteorological Instrumentation Requirements to Characterize Diffusion From Point Sources--A Workshop. Concord, MA: Environmental Research and Technology, 1980.

A methodological discussion of technical configurations utilized and recommended for field monitoring of effluent and atmospheric pollution. Useful for scientists from many disciplines.

745. Stokes, P.M. "pH-Related Changes in Attached Algal Communities of Softwater Lakes," in G.R. Hendrey (ed), Early Biotic Responses to Advancing Lake Acidification. Boston: Butterworth Publishers, 1984, pp. 43-61.

Report on a study in four lakes in Ontario in which plant populations were closely monitored along with lake chemistry. The experiment was an attempt to substantiate results obtained in simulated situations earlier and the results of this field trial were similar in many ways. Results indicate that a pH of 5-6 is critical for decline in aquatic plant biomass and speciation. Good scientific report but written so non-scientists can comprehend it.

746. Stopp, G.H. Jr. "The Regional Influence of Acid Rain on Nutrient Uptake in Agricultural Crops," in J. Hernandez (ed), A Healthy Economy in a Healthy Environment. Chapel Hill: University of North Carolina, 1983, pp. 161-168.

Preliminary report on field tests of acid rain effects on nitrogen-fixing bacteria on soybean roots. Unique because of controls on soil type.

747. ________. "Acid Rain and Agriculture: Immediate and Long Range Effects," in J.H. Baldwin (ed), Proceedings of the 12th Annual Conference of the National Association for Environmental Education. Troy, OH: National Association for Environmental Education, 1984, paper no. 58.

A general overview of research into the effects of acid rain on agricultural systems. Also includes inferential statements extending the research results into future situations. For a general, informed reader.

748. ________. "The Impact of Acidified Precipitation on Agricultural Crops," in A.B. Sacks (ed), Monographs in Environmental Education and Environmental Studies, Vol. 1. Columbus, OH: ERIC, 1984, pp. 173-182.

A good overview of what is known about the direct and indirect effects of acid rain on crop plants. No new research but a good compilation and synthesis from previous research by the author and by others.

749. Stoss, F.W. Economic Assessment of Acid Rain: A Bibliography. Rochester, NY: Center for Environmental Information, 1984.

Covers both published and unpublished works on one aspect of the acid rain controversy. More than 180 citations include conference papers and theses.

750. ________. Background Information on the Topic of Acid Rain, ERIC Bibliographic Series no. 1. Rochester, NY: Center for Environmental Information, 1984.

A short bibliography (7 pages) on acid rain. Not annotated or broken down by subject or area of interest.

751. Stottlemeyer, J.R. "The Neutralization of Acid Precipitation in Watershed Ecosystems of the Upper Peninsula of Michigan," in F.M. D'Itri (ed), Acid Precipitation: Effects on Ecological Systems. Ann Arbor, MI: Ann Arbor Science Publishing, 1982, pp. 261-275.

An examination of geologic neutralization of acid rain and snow in the Keeweenaw Peninsula and Isle Royale, Michigan region. Includes some data and analysis of seasonal precipitation and snowpack chemistry. Comprehensive report with good graphic and data presentations. Useful and applicable to other northern regions.

752. Streets, D.G., et al. "Selected Strategies to Reduce Acidic Deposition in the United States," Environmental Science and Technology, Vol. 17, no. 10, 1983, pp. 474A-485A.

Examines technological and regulatory attempts to slow acid rain. Really a shopping list of suggested methods.

753. Stubbs, H.S. Acid Rain Resources Directory. St. Paul, MN: Acid Rain Foundation, 1982.

A relatively incomplete compilation of persons, organizations, and agencies with an interest in acid rain research or control.

754. Summers, P.W. and D.M. Whelpdale. "Acid Precipitation in

Canada," Water, Air, and Soil Pollution, Vol. 6, 1976, pp. 39-44.

A complete early overview article on the reported and anticipated impact of acid precipitation in Canada. Gives insight into the background of policy development as well as scientific awareness.

755. Sze, N.D. and M.K.W. Ko. "Photochemistry of COS, CS_2, CH_3, SCH_3, and H_2S: Implications for the Atmospheric Sulfur Cycle," Atmospheric Environment, Vol. 14, 1980, pp. 1223-1239.

A very technical report on atmospheric chemical reactions that can affect the rate of sulfuric acid production downwind of effluent sites. Basic scientific data and analysis.

756. Szepesi, D.J. "Transmission of Sulfur Dioxide on Local, Regional and Continental Scales," Atmospheric Environment, Vol. 12, 1978, pp. 529-535.

A relatively early article outlining the transportation of pre-acidic chemicals through the atmosphere. Very useful in understanding the basic physics and chemistry of the production of acid rain.

757. Tamm, C.O. "Acid Precipitation: Biological Effects in Soil and on Forest Vegetation," Ambio, Vol. 5, No. 5-6, 1976, pp. 19-27.

A general article based more on inference from previous research than on actual acid rain tests. Good as a background reference.

758. Tanaka, S., et al. "Sulfur and Associated Elements and Acidity in Continental and Marine Rain from North Florida," Journal of Geophysical Research, Vol. 85, no. 8, 1980, pp. 4519-4526.

A thorough report on research into the chemistry of precipitation from continental vs marine source regions. Florida provides a unique situation for this type of comparative study. Marine precipitation could be considered "natural" while that formed over land anthropogenically altered. Excellent comparative data and analysis. Well written enough to be usable by all scientists.

759. Tanner, R.L., et al. "Acidity of Atmospheric Aerosols," Environmental Science and Technology, Vol. 15, no. 10, 1981, pp. 1150-1153.

A short look at specific chemical cycles in the atmosphere. Focuses on nitrogen and sulfur but can be used as early reading to understand the basic chemistry of acid rain and related phenomena.

760. Taubenfield, H.J. "Atmospheric Change, Politics, and World Law," Denver Journal of International Law and Policy, Vol. 10, no. 3, 1981, pp. 469-486.

A legal article concerned with the development of national and international environmental laws in reaction to increasing atmospheric pollution. Gives special attention to hypothetical situations involving transboundary pollution. Good source for policy students.

761. Taylor, J.E. and W.C. Leininger. "Monitoring Plant Community Changes Due to SO_2 Exposures," in The Bioenvironmental Impact of a Coal-fired Power Plant, EPA-600/3-78-021. Washington, DC: U.S. Government Printing Office, 1978, pp. 376-384.

A short but thorough evaluation of exposure techniques utilized in plant effects studies. Looks especially closely at the environmental conditions plants may experience and how the varied conditions affect plant and pollutant experiments.

762. Teknicron Research, Inc. Coal Resources and Sulfur Emission Regulations: A Summary of Eight Eastern and Midwestern States. Berkeley, CA: Teknicron Research, 1981.

More a listing of regulatory statutes for these states than a research report. Useful for comparative policy analysis or as background for policy or regulatory development.

763. Tetra Tech, Inc. The Integrated Lake-Watershed Acidification Study: Proceedings of the ILWAS Annual Review Conference, prepared for the Electric Power Research Institute (EA-2827). Washington, DC: EPRI, January, 1983.

Good collection of articles on statistical studies of drainage basins. The thrust is on the geological control of lacustrine acidity as is the interest of EPRI.

764. Theodore, L. et al. "Atmospheric Dispersion and Deposition of Particulates," in D.G. Nichols, et al. (eds), Energy and the Environment, Proceedings of the 5th National Conference. Dayton, OH: American Institute of Chemical Engineers, 1978, pp. 206-211.

A step-by-step development of an equation and calculational scheme for general approximation of atmospheric dispersion of pollutants from a single combustion source. Offers good insight into plume and transport modeling.

765. Thomas, F.W., et al. "Stacks--How High?," JAPCA, Vol. 13, no. 5, 1963, pp. 198-204.

One of the first articles focusing on the scientific debate about tall stacks. Very useful for an understanding of the development of policy designed to control local pollution problems but which led to national and international acid rain situations.

766. Thompson, R.J. "Design and Testing of a Prototype Rainwater Sampler/Analyzer," in U.S. E.P.A., Proceedings: National Symposium on Recent Advances in Pollutant Monitoring of Ambient Air and Stationary Sources, EPA-600/9-84-001. Washington, DC: U.S. Government Printing Office, 1983, pp. 290-299.

A report on the conceptual design of a new precipitation sampling device for use in acid rain research. Focuses on chemical changes that occur in transferring and analyzing rainwater, a major point of controversy in industry/government discussion. Good background for anyone planning field research.

767. Tollan, A. "Effects of Acid Precipitation on Aquatic and Terrestrial Ecosystems," in Restoration of Lakes and Inland Waters, EPA Report 440/5-81-010. Washington, DC: U.S. Environmental Protection Agency, 1980, pp. 438-445.

A rather general overview of acid rain effects on all ecosystems. No new data provided or new analysis attempted. Meant to serve as a base from which remedial solutions can be provided.

768. Tomlinson, G.H. "Air Pollutants and Forest Decline," Environmental Science and Technology, Vol. 17, no. 6, 1983, pp. 246A-256A.

An overview article of secondary research which attempts to present a general scenario for understanding forest effects of acid rain. No new information but a good synopsis.

769. Tong, E.Y., et al. "Regional and Local Aspects of Atmospheric Sulfates in the Northeastern Quadrant of the United States," in Proceedings of the Third Symposium on Turbulence, Diffusion and Air Quality. Boston: American Meteorological Society, 1976, pp. 139-145.

Fairly comprehensive report of measured and predicted sulfate levels in the atmosphere of the most acidified region in North America. Very useful data for temporal comparison.

770. Tonnessen, K.A. "Potential for Aquatic Ecosystem Acidification in the Sierra Nevada, California," in G.R. Hendrey (ed), Early Biotic Responses to Advancing Lake Acidification. Boston: Butterworth Publishers, 1984, pp. 147-169.

Report of a survey of the water quality of 40 western slope lakes in the Sierra Nevada range. Excellent baseline data provided as well as some analysis. Also comprehensive data on plankton population dynamics, correlated with lacustrine chemistry. An experimental microcosmic approach to sampling was utilized and it too is explained and analyzed as a research instrument. The particular geology of these lakes (vulnerable to acid effects) and the new approach to limnological sampling alone make this a worthwhile chapter; the data presented are also very useful.

771. Toribara, T.Y., et al. (eds). Polluted Rain. New York: Plenum Press, 1980.

A compilation of rather technical articles, on a wide range of topics from precipitation chemistry through effects studies and on to policy analysis. This work is the result of the 12th Rochester International Conference on Environmental Toxicity, held in 1979. A good general source for scientific data on acid rain and related problems.

772. Treshow, M. Whatever Happened to Fresh Air? Salt Lake City, UT: University of Utah Press, 1971.

A general text on air pollution but there are chapters on both acidic and pre-acidic deposition. There is also a very useful chapter on the effects of air pollution on agriculture. For a general informed reader.

773. Troiana, J. and E.J. Butterfield. "Effects of Simulated Acidic Rain on Retention of Pesticides on Leaf Surfaces," Phytopathology, Vol. 74, no. 11, 1984, pp. 1377-1380.

A research report on the effects of acid rain on fungicides on potato and bean plants in a synthetic situation. In general, the rain increases the rate of wash-off significantly. The real effects are found to be on the initial precipitation event. Of use to horticulturists and botanists.

774. Tukey, H.B., Jr. and J.V. Morgan. "Injury to Foliage and Its Effect Upon the Leaching of Nutrients From the Aboveground Parts of Plants," Plant Physiology, Vol. 16, 1963, pp. 557-565.

Not specific to acid rain effects but a very useful article for background information on some processes that are involved with acid rain effects on plants.

775. Tumlir, J. "Pollution Control and the Theory of Trade," in I. Walker (ed), Studies in International Environmental Economics. New York: Wiley-Interscience, 1976, pp. 110-116.

Not specific to acid rain but a general theoretical treatise on the interaction of government and business in pollution control. Case studies limited; more a discussion of general thoughts on this subject.

776. Turk, J.T. An Evaluation of Trends in the Acidity of Precipitation and the Related Acidification of Surface Water in North America, U.S. Geological Survey Supply Paper no. 2249. Washington, DC: U.S. Government Printing Office, 1983.

An excellent follow-up study on work done several years before by USGS. Provides comprehensive sets of data and good trend analysis. A very useful source of precipitation data.

777. Turner, J. and M.J. Lambert. "Sulfur Nutrition of Forests," in D.S. Shriner, et al. (eds), Atmospheric Sulfur Deposition. Ann Arbor, MI: Ann Arbor Scientific Publishing, 1980, pp. 321-333.

An in-depth examination of the sulfur cycle in forest ecology. Special focuses on soil ecology and the effects of different levels of acidic precipitation.

778. Tveite, B. and G. Abrahamsen. "Effects of Artificial Acid Rain on the Growth and Nutrient Status of Trees," in T.C. Hutchinson and M. Havas (eds), Effects of Acid Precipitation on Terrestrial Ecosystems. New York: Plenum Press, 1980, pp. 305-318.

Scientific report on experimental acidification of rain in a forest environment. Without a sufficient temporal sequence, this work remains preliminary.

779. Tyler, G. Ecological Effects of Acid Deposition, Report PM 1636,245. Stockholm: National Swedish Environmental Protection Board, 1983.

A comprehensive overview of environmental effects of acid rain, using a large collection of international data and research reports as background. One of the best synopses to data.

780. Tyler, G. "The Impact of Heavy Metal Pollution on Forests: A Case Study of Gusum, Sweden," Ambio, Vol. XIII, no. 1, 1984, pp. 18-24.

Results of a long-term study are presented for a region of intense local atmospheric pollution and acidic precipitation. Much of the effect and, as a result, research discussion, focuses on biomass conversion in the upper soil layers. Good information about forest soil ecology under acid conditions.

781. Tyree, S.Y., Jr. "Rainwater Acidity Measurement Problems," Atmospheric Environment, Vol. 5, 1981, pp. 57-60.

Short methodological statement about technical problems associated with collection and measurement of precipitation for pH testing.

782. Ulrich, B., et al. "Chemical Changes Due to Acid Precipitation in a Loess-derived Soil in Central Europe," Soil Science, Vol. 130, October, 1980, pp. 113-120.

A very scientific examination of ionic dynamics in a specific soil under acid precipitation. Good article for comparative purposes with work on other soils or in other precipitation regimes. The importance of loessal soils to world agriculture makes these especially critical data.

783. Ulrich, B. "Effects of Acid Deposition," in S. Beilke and A.J. Elshout (eds), Acid Deposition. Boston: Reidel Publishing, 1983, pp. 31-41.

Basically a general ecosystem evaluation. Forest soils, vegetation, and microbial activity are all examined under the impact of acidic deposition. No treatment of water resources but a complete analysis of other portions of the forest ecology.

784. Underwood, J.K. Acidic Precipitation in Nova Scotia. Halifax, Nova Scotia: Department of the Environment, 1981.

A good overview of acid precipitation in the maritime region, one of the most heavily impacted regions in North America. For an informed lay reader.

785. United States Department of Energy. Emission and Efficiency Performances of Industrial Coal Stoker Fired Boilers, DOE/ET/10386-TI. Washington, DC: U.S. Government Printing Office, 1981.

A set of test results on the combustion efficiency of different types of coal-burning equipment used in industrial settings. Baseline technological information.

786. _______. Acid Rain Information Book. Park Ridge, NJ: Noyes Data Corporation, 1982.

Public-oriented fact book. An almanac of information but not scientifically oriented or documented.

787. _______. Acid Precipitation: A Bibliography, DOE/TIC-3399. Washington, DC: U.S. Government Printing Office, 1983.

A large compendium of bibliographic entries. Annotated but difficult to use because of its computer-generated text; poor printing quality also. Very heavily oriented toward governmental reports but useful nonetheless.

788. United States Department of the Interior. Clean Power Generation from Coal, R&D Report no. 84. Washington, DC: U.S. Government Printing Office, 1973.

A useful examination of the technology needed and available to allow use of coal in power plants without increasing air pollution. This is part of the series of federal reports that recommend "tall stacks" policies and the wide use of scrubbers on smokestacks and which ultimately led to the national problem of acid rain. Very useful for policy students.

789. United States Energy Information Administration. Cost and Quality of Fuels for Electric Utility Plants, 1982 Annual Report, DOE/EIA-0191(82). Washington, DC: U.S. Department of Energy, 1983.

A compendium of data on fuel supplies utilized by electric generators in the United States. The data on sulfur content of coal are specifically useful for acid rain study.

790. _______. 1982 Annual Energy Review, DOE/EIA-0384(82). Washington, DC: U.S. Department of Energy, 1983.

A very useful compilation of information about a wide range of energy-related matters. Good source of data and some analysis of energy-related pollution problems, including coal consumption by industry.

791. _______. Electric Power Annual Report, 1983, DOE/EIA-0348(83). Washington, DC: U.S. Department of Energy, 1984.

An annual update of the report referenced above.

792. United States Environmental Protection Agency. Air Quality Criteria for Sulfur Oxides, EPA 83-73-033. Washington, DC: U.S. Environmental Protection Agency, 1973.

Basic policy document outlining effluent levels for sulfur compounds. Basic to the understanding of development of acid rain problem and "tall stacks."

793. _______. Popex--Ranking Air Pollution Sources by Population Exposure, EPA-600/2-76-063. Washington, DC: U.S. Government Printing Office, 1976.

An attempt to develop a model that will predict human effects of air pollution on a regional and national level. The study is wide-ranging, very general. Only portions are related to acid rain.

794. _______. Near-Surface Air Parcel Trajectories--St. Louis, 1975, EPA-600/3-77-123. Washington, DC: U.S. Government Printing Office, 1977.

A comprehensive report of adaptation of the St. Louis Regional Air Monitoring System to air parcel prediction. Important for predicting the movement of specific polluted air parcels. This study extends beyond local or regional effects and is important for understanding pollution transport on a regional and national scale. Rather scientific.

795. _______. National Air Pollution Emission Estimates, 1940-1976, EPA-450/1-78-003. Washington, DC: U.S. Environmental Protection Agency, 1978.

An excellent source of raw data on atmospheric chemicals in the U.S. Both the amount of data and the longevity of the collection period make this a very valuable source of information.

796. _______. The Bioenvironmental Impact of a Coal-fired Power Plant, EPA-600/3-78-021. Washington, DC: U.S. Government Printing Office, 1978.

A collection of research reports on investigations at Colstrip, Montana around a coal-fired power plant. Includes atmospheric and environmental studies. A comprehensive collection, very useful for scientist and non-scientist alike.

797. _______. Environmental Effects of Increased Coal Utilization: Ecological Effects of Gaseous Emissions from Coal Combustion, Report EPA-600/7-78-108. Corvallis, OR: Environmental Research Laboratory, 1978.

A large compilation of data on a national scale and short analyses of projects supported by EPA across the country. A good reference work.

798. _______. Air Pollution and Your Health, OPA 54/8. Washington, DC: U.S. Government Printing Office, March, 1979.

A general discussion of potential health effects from air pollution; only a small bit on acid rain. For a general public audience.

799. _______. Research Summary: Acid Rain, EPA-600/8-79-028. Washington, DC: U.S. Government Printing Office, October, 1979.

A good collection of abstracts from research projects on acid rain. Wide range of topics covered but limited to federally supported projects.

800. _______. Methods Development for Assessing Air Pollution

Control Benefits, EPA-600/5-79-001d. Washington, DC: U.S. Government Printing Office, 1979.

Primarily an economic analysis of the effects of air pollution regulation. Provides a justification of current and pending EPA policies and procedures. Useful for policy analysis.

801. _______. Symposium on Energy and Human Health: Human Costs of Electric Power Generation. Washington, DC: U.S. Government Printing Office, 1980.

A collection of presentations on health-related aspects of air pollution from power plants. Not all specifically related to acid rain but this can provide insight into the overall problem of combustion effects.

802. _______. Acid Rain, EPA-600/9-79-036. Washington, DC: U.S. Government Printing Office, 1980.

A general overview of the processes and problems associated with acid rain. No new data or analysis. Written for a general public but can be useful as a starting point for background information.

803. _______. National Air Pollutant Emission Estimates, 1940-1980, EPA 450/4-82-001. Washington, DC: U.S. Government Printing Office, 1982.

A large collection of primarily tabular data on air pollution in the U.S. Provides extensive data on all forms of pollution, including those related to acid rain. Excellent source, especially important because of the longevity of the data.

804. _______. Air Quality Criteria for Oxides of Nitrogen, EPA-600/8-82-026. Washington, DC: U.S. Government Printing Office, 1982.

Provides chapters on all aspects of the nitrogen cycle from combustion to recycling. Only one chapter dedicated to the acid rain phenomenon but that information is complete and well presented. Rather scientific but written so a non-scientific audience can understand it.

805. _______. Proceedings: National Symposium on Recent Advances in Pollutant Monitoring of Ambient Air and Stationary Sources, EPA-600/9-84-001. Washington, DC: U.S. Government Printing Office, 1983.

A large collection of articles on various aspects of air pollution. Much of this collection deals with non-acid rain related subjects but several chapters are directly on subject. A good reference book.

806. _______. National Air Pollutants Emission Estimates, 1940-1982, EPA-450/4-83-024. Washington, DC: U.S. Government Printing Office, 1984.

An update of the previous publication to include data since 1980. Also some refinement of the old data when possible. A comprehensive listing of information on pollution levels in the United States.

807. _______. The Acidic Deposition Phenomenon and Its Effects: Critical Assessment Review Papers, EPA-600/8-83-016. Washington, DC: U.S. Government Printing Office, 1984.

A large collection of articles on almost every aspect of acidic deposition. Volume 1 focuses on the atmospheric sciences, pollution production, and transport. Volume 2 focuses on environmental effects, including water resources, soil, forests, and agriculture. An excellent overall source of data and analysis. Recommended as one of the most comprehensive and current sources available.

808. United States Fish and Wildlife Service. Potential Impacts of Low pH on Fish and Fish Populations, FWS/OBS-80/40.2. Washington, DC: U.S. Government Printing Office, 1983.

Modeled inferences of national effects on fish resources in lakes and streams based on published research reports. No new information but interesting extrapolations of what was known at the time.

809. _______. Effects of Acid Precipitation on Aquatic Resources, FWS/OBS-80/40.12. Washington, DC: U.S. Government Printing Office, 1983.

A compilation of information from government-supported research into reactions of fisheries to acid rain. Good reference document for scientist and informed non-scientist alike.

810. _______. Liming of Acidified Waters: A Review of Methods and Effects on Aquatic Ecosystems, FWS/OBS-80/40.13. Washington, DC: U.S. Government Printing Office, 1983.

A good compilation of studies on lacustrine liming to counteract acidification. While the results are not very positive, they are important and necessary to applied study of acid rain.

811. United States General Accounting Office. The Debate Over Acid Precipitation: Opposing Views--Status of Research, EMD-81-131. Gaithersburg, MD: U.S. G.A.O., 1981.

No new research or insight into the subject, just a bureaucratic overview of acid rain research. Only use might be for policy-oriented study considering the effect GAO has on federal policy.

812. ________. An Analysis of Issues Concerning "Acid Rain," GAO/RCED-85-13. Gaithersburg, MD: U.S. General Accounting Office, 1984.

A rather general overview of the state-of-knowledge of the effects of acid rain on a wide spectrum of segments of the environment. No new knowledge and, except for the government documents referenced, the research drawn from is rather old. Of use to policy analysts because of the direct impact GAO has on national legislation and federal attitudes but of very limited scientific utility. The gist of the document is that there is not enough real evidence on the effects of acid rain to develop meaningful regulations for its control.

813. ________. Cost-Benefit Analysis Can Be Useful in Assessing Environmental Regulations, Despite Limitations, GAO/RCED-84-62. Gaithersburg, MD: General Accounting Office, 1984.

An econometric treatise on methods used to determine trade-offs in regulating environmental problems. Aimed at acid-rain regulation specifically but useful for general environmental economics also.

814. United States House of Representatives. Acid Rain: Implications for Fossil Energy R&D, Hearing before the House Committee on Science and Technology, September 20, 1983. Washington, DC: U.S. Government Printing Office, 1983.

A transcript of an excellent session of the Committee on Science and Technology which focuses on the national and international need for specific research on acid rain. The effects of this research and the acid rain problem on the future use of coal in the U.S. is discussed broadly. Of great value to economists and policy students.

815. United States Interagency Task Force on Acid Precipitation. Annual Report, 1983, to the President and Congress. Washington, DC: U.S. Government Printing Office, 1984.

A compilation of information on acid rain as of 1983, in a nontechnical format. For a very general reader but it does include usable graphics and data taken from other sources. No new data or research presented but an acceptable condensation of technical and scientific work by others.

816. United States Office of Technology Assessment. Acid Rain and Transported Air Pollutants, Implications for Public Policy, OTA-0-204. Washington, DC: U.S. Government Printing Office, 1984.

Primarily a policy study; only minimal use of scientific data or analysis. Several regulatory scenarios are discussed and/or proposed and analyzed with an eye to economic and technical effects. Of interest primarily to students of public policy.

817. United States Senate Committee on Energy and Natural Resources. Sulphur Oxides Control Technology in Japan, Washington, DC: EPA Interagency Task Force, June 30, 1978.

A very general presentation of an analysis of Japanese attempts to control effluent from coal combustion.

818. United States Senate Committee on Environment and Public Works. Acid Rain: A Technical Inquiry, 97-H53. Washington, DC: U.S. Government Printing Office, 1982.

A general, rather bureaucratic document meant to serve as a briefing for legislators concerned with environmental policy. Of special interest is an appendix on the costs of controlling acid rain that will be of interest to economists and to policy students.

819. Unsworth, M.H. and D.P. Ormrod (eds). Effects of Gaseous Air Pollution in Agriculture and Horticulture. London: Butterworth Scientific, 1982.

A collection of international articles on air pollution and agriculture. Not specific to acid rain or to acid-producing pollutants but several articles address this problem directly. Good comparative data and analysis of interest to ecologists as well as agricultural scientists.

820. Uthe, E.E. and W.E. Wilson. "Lidar Observations of the Density and Behavior of the Labadie Power Plant Plume," Atmospheric Environment, Vol. 13, 1979, pp. 1395-1412.

Rather technical report of observations of the daily physics and chemistry of a major effluent source. Data are useful for limited comparisons.

821. Uthe, E.E., et al. "Lidar Observations of the Diurnal Behavior of the Cumberland Power Plant Plume," JAPCA, Vol. 30, no. 8, 1980, pp. 889-893.

Technical research report on daily dynamics of a large point-source for pre-acidic effluent. Of use for data and for understanding this specific portion of the acid rain cycle.

822. Varshney, C.K. and L.S. Dochinger. "Acid Rain: An Emerging Environmental Problem," Current Science, Vol. 48, no. 8, 1979, pp. 337-340.

A rather general article simply enumerating the effects acid rain is having on various portions of the environment. No new data but a synopsis of what was known at the time of publication.

823. Vermeulen, A.J. "The Acidic Precipitation Phenomenon: A Study of This Phenomenon and of a Relationship Between the Acid Content of Precipitation and the Emission of Sulfur Dioxide and Nitrogen Oxides in the Netherlands," in T.Y. Toribara, et al. (eds), Polluted Rain. New York: Plenum Press, 1980, pp. 7-60.

A thorough examination of the changes in precipitation pH in the Netherlands over time, wich special concern for the period immediately surrounding 1966. While atmospheric chemistry is explained, there is considerable concern with the changes in Dutch fuel consumption, correlating acidity with oil, coal and natural gas usage. The relationship established between acidity and each fuel is especially useful and transferable.

824. Volchok, H.L. "Atmospheric Deposition of Man-made Radioactivity," in T.Y. Toribara, et al. (eds). Polluted Rain. New York: Plenum Press, 1980, pp. 435-448.

While not directly focused on acid rain, this article describes a governmental precipitation collection and analysis network, established to measure radioactivity, that can have methodological implications for acid rain researchers.

825. Vozzo, S.F. International Directory of Acid Deposition Researchers. St. Paul, MN: Acid Rain Foundation, 1983.

A partial listing of people who have been or are active in acid-rain-related research and publication. Incomplete and dated.

826. Waddell, T.E. The Economic Effects of Sulfur Oxide Air Pollution From Point Sources on Vegetation and Environment, Policy Paper no. 70-116. Washington, DC: National Air Pollution Control Administration, 1970.

A general examination of research into the effects of SO on the environment. Much of the data is inferred; the early date precludes having access to much experimental data. Written in a non-technical form for use by policy makers. The focus is on point-source pollution, something most acid-rain research is not concerned with.

827. Wagner, G.H. and K.F. Steele. Nutrients and Acid in the Rain and Dry Fallout at Fayetteville, Arkansas (1980-1982), Project A-051-ARK. Little Rock, AR: U.S. Department of the Interior, 1983.

A thorough regional report on monitoring of wet and dry deposition in Arkansas. The lengthy study period and the comparative data sets (wet and dry) make this a very valuable source. The location, mid-south U.S., an area in which very little research on acid rain has been done, also makes this report important. Good data sets and analysis. Useful for all acid rain researchers.

828. Walker, J.T. Characteristics of Rain in the Georgia Piedmont and Effects of Acidified Water on Crop and Ornamental Plants, University of Georgia Agricultural Experiment Station Research Bulletin 283. Athens, GA: University of Georgia, 1982.

A good regional report. Useful sets of data both on precipitation chemistry and on experiments with crop and non-crop plants. The mixed results on crop plants is particularly useful and interesting. Good source for agronomists and botanists.

829. Warburton, J.A. "The Chemistry of Precipitation in Relation to Precipitation Type," in M.M. Benarie (ed), Atmospheric Pollution 1982. New York: Elsevier Publishing Co., 1982, pp. 379-386.

Report on research on the effects of specific precipitation types on precipitation chemistry. Particle size, type of ice crystal formed, rates of Brownian capture all examined as controls of pollution capacity. Rather technical and for atmospheric scientists only.

830. Wark, K. and C.F. Warner. Air Pollution: Its Origin and Control. New York: Harper and Row, 1976.

A comprehensive overview of all aspects of air pollution. Causes, effects, technological matters, regulation, and measurement are all treated in this rather technical text. Acid rain treated as simply one aspect of air pollution.

831. Watt, W.W. "Acidification and Other Chemical Changes in Halifax County Lakes After 21 Years," Limnology and Oceanography, Vol. 24, no. 6, 1979, pp. 1154-1161.

A comprehensive limnological study of changes in lake chemistry over an extended study period. Of use for comparative analysis and as a baseline of data for comparable studies.

832. Weller, P. Acid Rain: The Silent Crisis. Kitchener, Ont.: Waterloo Public Interest Research Group, 1980.

An interesting general text written from a concerned citizen's standpoint. An advocacy document.

833. Wesely, M.L. and B.B. Hicks. "Dry Deposition and Emission of Small Particles at the Surface of the Earth," Proceedings of the Fourth Symposium on Turbulence, Diffusion and Air Pollution. Boston: American Meteorological Society, 1979, pp. 510-513.

A rather technical analysis of the processes involved with particle selection in dry deposition of pollutants. Very useful for meteorologists and atmospheric chemists.

834. Wesler, N. "Air Pollution and the Projected Consumption of Fossil Fuels for Purposes Other Than Internal Combustion Engines," in A.J. Van Tassel (ed), Environmental Side Effects of Rising Industrial Output. Lexington, MA: Heath Lexington Books, 1970, pp. 165-190.

A general look at the potential levels of production of NO_x and SO_x with rising industrial use of coal. Projects levels of use and pollution to the year 2,000. Data and projections on all industrial and residential uses. A good source of projection data.

835. West, S. "Acid From Heaven (rain isn't right any more and where it falls, it creates a nasty mess--environmentally and politically)," Science News, Vol. 117, 1980, pp. 76-78.

A journalistic overview of environmental and policy-related reactions to acid rain in North America.

836. Wetstone, G. "Air Pollution Control Laws in North America and the Problems of Acid Rain and Snow," Environmental Law Reporter, Vol. 10, 1980, pp. 50001-50020.

A comprehensive and clearly written overview of the development of and the further need for regulation of air pollution. The author is an "environmental advocate" but his treatment of this subject is objective.

837. ________. "Acid Precipitation: The Need for a New Regulatory Approach," Environment, Vol. 22, no. 5, 1980, pp. 9-14.

An innovative article by a nationally known environmental lawyer who recognizes that the present regulatory system has failed and that more of the same will not work. Good for policy students and scientists alike.

838. ________ and S. Foster. "Acid Precipitation: What is it Doing to Our Forests?," Environment, Vol. 25, no. 4, 1981, pp. 10-12.

An article written for a general, non-scientific audience. Not particularly enlightening or insightful. No new data or analysis but some new thoughts as legal and policy concerns are raised.

839. Wetstone, G. "Air Pollution Control Laws in North America and the Problem of Acid Rain," in G.H. Stopp, Jr., Environmental Management. Washington, DC: Gemini Press, 1981, pp. 60-108.

A good general discussion of the development of U.S. environmental policy as it affects air pollution and its by-product, acid rain and snow. Well written.

840. Whatley, S. and M. Koziol (eds). Gaseous Air Pollutants and Plant Metabolism. Boston: Butterworth Science, 1984.

A thorough compilation of articles on many aspects of air pollution effect on plant processes. Not limited to acidic or acid-producing pollutants but does include several chapters specific to this subject. An up-to-date source of research and analysis.

841. Whelpdale, D.M. "Acidic Deposition," in Ecological Effects of Acid Precipitation, a report of a workshop held at Gatehouse-of-fleet, Galloway, U.K., September 4-7, 1978, pp. 12-16.

A very general overview of the acid deposition phenomenon. An introductory statement.

842. White, W.H., et al. "Formation and Transport of Secondary Air Pollutants: Ozone and Aerosols in the St. Louis Urban Plume," Science, Vol. 194, 1976, pp. 187-189.

A useful report of atmospheric effluent monitoring in a major U.S. urban region. Pre-tall stacks, localized phenomenon is analyzed but the physics and chemistry of these events is applicable regionally and nationally.

843. Whitehead, D.R., et al. "Late Glacial and Postglacial pH Changes in Adirondack Lakes," Bulletin of the Ecological Society of America, Vol. 62, 1981, p. 154.

A short note on geological evidence of paleo-atmospheric conditions.

844. Wiegner, J.G., et al. "Species Composition of Fish Communities in Northern Wisconsin Lakes: Relation to pH," in G.R. Hendrey (ed), Early Biotic Responses to Advancing Lake Acidification. Boston: Butterworth Science Publishers, 1984, pp. 133-146.

A scientific comparison of 12 lakes with acidic or neutral pH levels. Fish speciation, fish population dynamics and productivity are all examined and analyzed. General results indicate that speciation is richer in less acid lakes. A good source of information for ecologists and for fishery scientists.

845. Wiener, J.G. "Acidic Precipitation: A Selected, Annotated Listing of Information Sources," Fishers, Vol. 8, no. 4, 1983, pp. 7-11.

A very short bibliography of material on acid rain. Good but limited.

846. Wiklander, L. "Leaching and Acidification of Soils," in M.J. Wood (ed), Ecological Effects of Acid Precipitation. Surrey: U.K.: Central Electricity Research Laboratories, 1979, pp. 46-54.

A synthetic analysis of ionic activity of acid precipitation as it percolates through the soil profile. No new research but a well-presented overview of the processes.

847. _______. "The Sensitivity of Soils to Acid Precipitation," in T.C. Hutchinson and M. Havas (eds), Effects of Acid Precipitation on Terrestrial Ecosystems. New York: Plenum Press, 1980, pp. 553-567.

A very scientific but often inferential chapter on the way in which the geology and ecology of soils change and are changed by acidic precipitation. Good overview article.

848. Wilson, B.D. "Nitrogen and Sulfur in Rainwater in New York," Journal of the American Society of Agronomy, Vol. 18, 1926, pp. 1108-1112.

A very early report of precipitation chemistry. These data are excellent for temporal analysis and lend themselves to linear comparison with later time periods and fossil fuel combustion data.

849. Winkler, E.M. "Natural Dust and Acid Rain," Water, Air and Soil Pollution, Vol. 6, no. 2, 1976, pp. 295-302.

A report, part from research, part from inference, on how natural atmospheric dust counteracts the production of acidic air. Certain types of dust turn carbonates into harmless gypsum. An interesting article for atmospheric scientists.

850. Winkler, P. "Deposition of Acid in Precipitation," in H.W. Georgii and J. Pankrath (eds), Deposition of Atmospheric Pollutants. Boston: Reidel Publishing, 1982, pp. 67-76.

A technical article that focuses on transport trajectory from point sources. Provides a comparative analysis of transport patterns and processes over land masses and over water bodies. Also demonstrates that the size of rain drop has a direct effect on acid deposition. Good source for atmospheric chemists and meteorologists.

851. _______. "Trend Development of Precipitation pH," in S. Beilke and A.J. Elshout (eds), Acid Deposition. Boston: Reidel Publishing, 1983, pp. 114-122.

Duplicates in many ways the work above by the same author but focuses more on overall European precipitation pH trends. His data indicate that, even though SO_2 is up, the precipitation is not increasingly acidic because a saturation point may have been reached. Very useful data and interesting inferences and theory development.

852. Wisniewski, J. and J.D. Kinsman. "An Overview of Acid Rain Monitoring Activities in North America," *Bulletin of the American Meteorological Society*, Vol. 63, 1982, pp. 598-618.

An excellent, comprehensive listing and analysis of the many precipitation monitoring programs in North America which focus on acid rain. Also looks at effects monitoring programs. Very useful to gain an understanding of the research networks which can provide data on acid rain. Also useful for policy study.

853. Wolt, J.D. and D.A. Lietzke. "The Influence of Anthropogenic Sulfur Inputs Upon Soil Properties in the Copper Basin Region of Tennessee," *Soil Science Society of America Journal*, Vol. 46, May/June, 1982, pp. 455-467.

An excellent study of a region similar in many ways to the Sudbury area of Canada. Presents modern data with considerable longevity to it.

854. Wood, J.M. "The Role of pH and Oxidation-Reduction Potentials in the Mobilization of Heavy Metals," in T.Y. Toribara, et al. (eds), *Polluted Rain*. New York: Plenum Press, 1980, pp. 223-236.

Another article examining the relationship between acid precipitation and heavy metal deposition and effects. This focuses on the accumulation of mercury in fish taken from acid-sensitive lakes. Rather technical physical chemistry but very useful for limnologists and ecologists.

855. Wood, T. and F.H. Bormann. "The Effects of an Artificial Mist Upon the Growth of Betula Allegheniensis," *Environmental Pollution*, Vol. 7, 1974, p. 259.

Basic research into ecological effects of acid rain. For scientific audiences only.

856. ________. "Increases in Foliar Leaching Caused by Acidification of an Artificial Mist," *Ambio*, Vol. 4, 1975, pp. 169-171.

Short research report on leaf study using simulated acid rain. Very useful for botanists and ecologists. Excellent write-up on physiological processes under acid stress.

857. Wright, R.F. and E.T. Gjessing. "Acid Precipitation: Changes in the Chemical Composition of Lakes," *Ambio*, Vol. 5, no. 5-6, 1976, pp. 231-234.

A short report on acidification of Swedish lakes. This is a good early source of data on lacustrine chemistry in a very sensitive region. Not much discussion or data on secondary effects of water acidification but a thorough examination of lake chemical processes.

858. Wright, R.F., et al. "Acidified Lake Districts of the World: A Comparison of Water Chemistry of Lakes in Southern Norway, Southern Sweden, Southwestern Scotland, the Adirondack Mountains of New York and Southeastern Ontario," in D. Drablos and A. Tollan (eds), Proceedings of the International Conference on the Ecological Impact of Acid Precipitation, Oslo: SNSF Project, 1980, pp. 377-379.

An excellent comparative article. No new data but a new presentation of old data with interesting comparisons and parallels across regions.

859. Wyzga, R. Environmental Impacts of Dispersed and Concentrated Siting of Coal-fired Power Plants, EA-3484. Palo Alto, CA: Electric Power Research Institute, 1984.

A thorough study of the differences in real and inferred environmental impact brought about by the two strategies of dispersing and concentrating coal-fired plants. There are negative and positive aspects to both presented along with some analysis of the real trade-offs involved with selection of either strategy. Good for policy analysis and for economists.

860. Yandle, B. "The Emerging Market in Air Pollution Rights," Regulation, July/August, 1978, pp. 21-29.

Written for a general audience but presents an interesting economic approach to citizen input into air pollution control.

861. Yue, G.K., et al. "A Mechanism for Hydrochloric Acid Production in a Cloud," Water, Air, and Soil Pollution, Vol. 6, 1976, pp. 277-294.

A comprehensive scientific analysis of the chemistry involved with one form of acid rain production. Very well written but technical.

862. Zeekijk, H. and C. Velds. "The Transport of Sulfur Dioxide Over a Long Distance," Atmospheric Environment, Vol. 7, 1973, pp. 849-862.

A technical article written "pre-acid rain" but concerned with one of the atmospheric processes that is basic to acid rain production and distribution. Provides an excellent model of the movement of sulfur through the atmosphere.

863. Zimmerman, A.P. and H.H. Harvey. Sensitivity to Acidification of Waters of Ontario and Neighboring States, Final Report for Ontario Hydrology. Toronto: University of Toronto, 1980.

A comprehensive, scientific study of the potential impact of acid rain on Canadian lakes and streams. Recommended for scientists and non-scientists alike.

864. _______. Air Pollution Across National Boundaries--The Impact on the Environment of Sulfur in Air and Precipitation. Stockholm: Sweden's Royal Ministry for Foreign Affairs, 1971.

Essentially a case study of acidic precipitation in Sweden prepared by the Swedish government for the 1972 United Nations Conference on the Human Environment. Rather general but comprehensive in the range of subjects covered. A good case study of a national problem.

865. _______. Acid Precipitation and Its Effects in Norway, Oslo: Ministry of Environment, 1974.

A general case study of overall acid precipitation problems in Norway. Similar to the report for Sweden mentioned above; also prepared for the U.N. conference but published subsequent to that meeting.

866. _______. "Report of the International Conference on the Effects of Acid Precipitation in Telemark, Norway," Ambio, Vol. 5, 1976, pp. 200-252.

A short, journalistic report. Simply outlines the highlights of this important meeting.

867. _______. The MAP3S Precipitation Chemistry Network: First Periodic Summary Report (September 1976-June 1977). Richland, WA: Battelle Pacific Northwest Laboratories, 1977.

A comprehensive summary of the data collected and of the data collection procedures and network established under MAP3S. A landmark reference both for the data presented and for the network analysis.

868. _______. The Depression of pH in Lakes and Streams in Central Ontario During Snowmelt. Ottawa: Ontario Ministry of the Environment, 1979.

A comprehensive research report on snowmelt and the periodic input of acids into lakes and streams. Excellent source of data and methodological information.

869. _______. Acidic Precipitation in South-Central Ontario: Analysis of Source Regions Using Air Parcel Trajectories. Ottawa: Ontario Ministry of the Environment, 1980.

An inferential analysis for a regional impact study. Relatively useful for a comparative study of transportation models.

870. _______. The Long Range Transport of Air Pollutants, Second Report of the United States-Canada Research Consultation Group. Ottawa: Canadian Ministry of the Environment, 1980.

A bureaucratic attempt at scientific analysis. More useful for policy students than for scientists.

871. _______. "The Acid Earth," Harrowsmith, Vol. 68, 1980, pp. 32-93.

A general article for public consumption. No new information or insight.

872. _______. The Case Against Acid Rain. Toronto: Ontario Ministry of the Environment, 1980.

Produced for public consumption, this does give an overview of the whole range of environmental effects of acid rain. Gives some insight into the development of a policy attitude by a regional governmental bureaucracy.

873. _______. U.S.-Canada Memorandum of Intent on Transboundary Air Pollution--Emissions, Costs and Engineering Assessment, NTIS-PB-174-724. Washington, DC: U.S. Environmental Protection Agency, 1981.

A comprehensive study of proposed technological implementations designed to retard the emission of pre-acidic pollutants from major pollution sites in the U.S. and Canada. Provides not only clear and lengthy (177 pp.) analysis of available and proposed technological developments but estimates of costs/benefits related to each technological level.

874. _______. U.S.-Canada Memorandum of Intent on Transboundary Air Pollution--Atmospheric Modelling, NTIS-PB81-173064. Washington, DC: U.S. Environmental Protection Agency, 1981.

A lengthy (184 pp.) report on scientific attempts to determine predictive models for atmospheric transport of air pollution and the subsequent creation of acid precipitation across the U.S./Canada frontier. No new information but a clear, non-technical presentation of accepted models along with coherent analysis. Very useful for atmospheric scientists as well as policy analysts.

875. _______. U.S.-Canada Memorandum of Intent on Transboundary Air Pollution--Impact Assessment, NTIS-PB81-179376. Washington, DC: U.S. Environmental Protection Agency, 1981.

An international report on the effects of acidic precipitation on portions of the U.S. and Canada. Provides no new data but compiles conveniently and analyzes data from a comprehensive list of sources. Excellent for general scientific background or policy students.

876. _______. Memorandum of Intent on Transboundary Air Pollution, Group 2 Interim Report of the United States-Canada Work Group. Ottawa: Canadian Ministry of the Environment, 1981.

Really a compendium of scientific outlines of research and analysis on the various aspects of acid rain. Very useful for scientists and policy analysts.

877. _______. The Costs and Benefits of Sulphur Oxide Control--A Methodological Study. Paris: Organization for Economic Cooperation and Development, 1981.

An econometric analysis of various scenarios designed to control the chemical cause of acid rain. No new insights but the international range of this text makes it unique.

878. _______. Forest Damage Due to Air Pollution. Bonn: Federal Republic of Germany Minister for Nutrition, Agriculture and Forestry, 1981.

A thorough assemblange of research reports on acid rain effects in forest ecosystems. This report includes other pollutants as well but the major focus is on acid rain and pre-acid rain chemicals. Paints a very negative picture of forest damage but reports it on a regional level and does not provide analysis or data on the specific physiological effects.

879. _______. The Effects of Acid Deposition on Environmental Amenities. Toronto: Ontario Ministry of the Environment, 1982.

Written for the general public. No new information but a well-packaged presentation of synthetic analyses on the whole range of acid rain effects.

880. _______. "A Blow to Acid Rain," Science News, Vol. 117, no. 26, 1982, p. 407.

Journalistic report on policy developments. No new information or analysis but a cursory look at public opinion and policy development.

881. _______. 1982 Stockholm Conference on Acidification of the Environment. Stockholm: Swedish Ministry of Agriculture, 1982.

A collection of papers which focus on all aspects of acidification, from pollution production to socio-economic effects of pollution control. A very useful collection of research reports on Swedish environmental problems.

882. _______. Acidification Today and Tomorrow. Stockholm: Swedish Ministry of Agriculture, 1982.

A very lengthy (232 pp.) report on the overall acidification of the Swedish and northern European environment by air pollution. Thorough presentation of data on forest and water resources as well as analysis and predictive models of future effects.

883. _______. Acidification: A Boundless Threat to Our Environment. Stockholm: Swedish Ministry of Agriculture, 1983.

A rather short report on the international aspects of acidification in northern Europe focusing on Sweden and the Scandinavian peninsula. Provides no new information and only general synopses of the various environmental problems created by acid rain.

884. _______. Acid Rain--A Review of the Phenomenon in the EEC and Europe. London: Environmental Resources Limited, 1983.

A general report prepared for the Commission of the European Communities which highlights the various local, regional, and international environmental problems created by acid rain. This also examines the economic implications of both pollution control implementation and continued lack of control.

885. _______. Acid Deposition in North America--A Review of the Documents Prepared Under the Memorandum of Intent Between Canada and the United States of America on Transboundary Air Pollution. Ottawa: Royal Society of Canada, 1983.

A two-volume report on scientific and policy related research and reports prepared under the 1980 Transboundary Agreement. Some analysis of research needs and trends provided but no new data or analysis forthcoming.

886. _______. Economic Commission for European Executive Body for the Convention on Long-Range Transboundary Air Pollution. Geneva: United Nations Economic and Social Council, 1983.

A short report of economic concerns related to air pollution and transboundary problems of effects and control regulation. More a policy background document than a research or analytical report.

NADP/NTN DEPOSITION MONITORING PROGRAM . . . .

As a response to the need to understand the impact of the deposition of atmospheric chemicals on agricultural resources, the North Central region of the State Agricultural Experiment Stations established NC-141, a program to conduct research on the effects of atmospheric deposition. This program developed into an interregional project called IR-7 and eventually became the National Atmospheric Deposition Program (NADP).

NADP now consists of an interdisciplinary team of more than 100 research scientists and has a network of more than 150 atmospheric deposition monitoring sites across the United States. Federal and regional agencies that participate in NADP include: The U.S. Department of State, the State Agricultural Experiment Stations (SAES), the U.S. Department of Agriculture (USDA), the U.S. Department of Interior (subagencies USGS, NPS, BLM), the U.S. Environmental Protection Agency (EPA), the Tennessee Valley Authority (TVA), the National Aeronautics and Space Administration (NASA), the National Oceanic and Atmospheric Administration (NOAA), the Electric Power Research Institute (EPRI), and the Atmospheric Environment Service of Canada.

In 1982 the U.S. Congress established the National Acid Precipitation Assessment Program (NAPAP) to study the problem of acid deposition on a national level. As part of its charge, NAPAP was to establish a regional deposition monitoring network similar to that already in place under NADP supervision. This new network, designated the National Trends network (NTN), was intended to have 150 collection sites and, since NADP was already in place, NTN simply selected existing NADP sites and contracted with that organization to provide deposition data.

Every site is equipped with the same set of measuring devices. Each has a wet/dry deposition collector (Aerochem Metrics Model 301), pH and conductivity meters and a recording rain gauge. Weekly samples of precipitation are analyzed by the Illinois State Water Survey Central Analytical Laboratory. At present, analyses are conducted on each sample for SO_4, NO_3, PO_4, NH_4, K, Ca, Mg, and pH. In addition, tests for electrical conductivity occur regularly, and tests for heavy metals, especially Fe, Zn, Hg, Pb, and Cd, may be performed upon request. All analyses are overseen by the NADP Quality Assurance Steering Committee with the cooperation and support of the Department of Energy, the Environmental Protection Agency and the U.S. Geological Survey.

NADP data are available to all members of NADP and NTN and any portion of the data will be provided upon request for any purpose of analysis or interpretation. Anyone wishing to obtain NADP data or to gain an understanding of the workings of the NADP or NTN monitoring systems should contact:

Mr. J.H. Gibson
NADP Coordinator
Natural Resource Ecology Laboratory
Colorado State University
Fort Collins, Colorado 80523
Telephone: 303-491-1978

The NADP is governed by a Technical Committee made up of scientists who work with or have an interest in acidic deposition. Four subcommittees have been organized to provide technical advice and direction for all portions of the NADP program. NADP holds an annual meeting at which both policy matters and scientific papers are discussed. These meetings are open to anyone who wishes to attend and participate.

On the following pages are provided a list of officers in NADP and their addresses and a listing of the official NADP (and thus NTN) collection sites.

NATIONAL ATMOSPHERIC DEPOSITION PROGRAM 1983-84 . . . .

Executive Committee

Program Chairman	William W. McFee	Agronomy Department--LILY Purdue University West Lafayette, IN 47907 (317) 494-8044
Program Vice Chairman	David S. Shriner	Environmental Sciences Division Oak Ridge National Laboratory Oak Ridge, TN 37830 (615) 574-7356
Program Secretary	Stephen A. Norton	Department of Geological Sciences University of Maine at Orono Orono, ME 04469 (207) 581-2156
Past Program Chairman	Ellis B. Cowling	School of Forest Resources North Carolina State University Raleigh, NC 27650 (919) 737-2883

Committee Chairmen

Site Committee	John K. Robertson	Science Research Laboratory U.S. Military Academy West Point, NY 10996 (914) 938-3739
Methods Committee	Donald C. Bogen	U.S.D.O.E. E.M.L. 376 Hudson Street New York, NY 10014 (212) 620-3637
Data Committee	Steven E. Lindberg	Oak Ridge National Laboratory PO Box X Oak Ridge, TN 37830 (615) 574-7857

Effects Committee	J.M. Kelly	Tennessee Valley Authority TVA/ORNL Watershed Study Program Building 1505, R. 338, ORNL Oak Ridge, TN 37830 (615) 574-7815

Subcommittees

Effects Research Committee

Chairman	J.M. Kelly	Tennessee Valley Authority TVA/ORNL Watershed Study Program Building 1505, Rm. 338, ORNL Oak Ridge, TN 37830 (615) 574-7815
Chairman-Elect	Dudley J. Raynal	College of Environmental Science and Forestry State University of New York Syracuse, NY 13210 (314) 470-6782
Vice Chairman	Jim Perry	Department of Forest Resources 110 Green Hall University of Minnesota St. Paul, MN 55108 (612) 373-0846

Quality Assurance Steering Committee

Chairman	Peter Finkelstein	U.S. E.P.A. MD-56 Research Triangle Park, NC 27711 (919) 541-2347

EFFECTS RESEARCH COMMITTEE WORKING GROUPS

Aquatics	E.S. Verry	USDA Forest Service 1831 Highway 169 East Grand Rapids, MN 55744 (218) 326-8571
Field and Horticultural Crops	Wayne Banwart	S-512 Turner Hall University of Illinois 1102 South Goodwin Urbana, IL 61801 (217) 333-9470

Forestry	James B. Hart	Department of Forestry Michigan State University East Lansing, MI 48824 (517) 355-9528
Materials	Raymond Herrmann	National Park Service 107C Natural Resources Colorado State University Fort Collins, CO 80523 (303) 491-7473

Network Site Criteria and Standards Committee

Chairman	John K. Robertson	Science Research Laboratory U.S. Military Academy West Point, NY 10996 (914) 938-3739
Vice Chairman	Jerry T. Walker	Department of Plant Pathology University of Georgia Georgia Station Experiment, GA 30212 (404) 228-7202
Secretary	Malcolm E. Still	Atmospheric Environment Service 4905 Dufferin Street Downsview, Ontario M3H 5T4 CANADA (416) 667-4988

Methods Development and Quality Assurance Committee

Chairman	Donald C. Bogen	U.S.D.O.E. E.M.L. 376 Hudson Street New York, NY 10014 (212) 620-3637
Vice Chairman	G.M. Aubertin	Department of Forestry Southern Illinois University Carbondale, IL 62901 (618) 453-3341

Data Management and Analysis Committee

Chairman	Steven E. Lindberg	Oak Ridge National Laboratory PO Box X Oak Ridge, TN 37830 (615) 574-7857

Vice Chairman	James A. Lynch	Pennsylvania State University 311 Forest Resources Lab University Park, PA 16802 (814) 865-8830

OTHER PROGRAM OFFICERS

Administrative Advisors	F.A. Wood	Agricultural Experiment Station University of Florida 1022 McCarty Hall Gainesville, FL 32611 (904) 392-1784
	W. Lamar Harris	Agricultural Experiment Station University of Maryland College Park, MD 20742 (301) 454-3707
	Lee A. Bulla, Jr.	Agricultural Experiment Station University of Idaho Moscow, ID 83843 (208) 885-6682
	Keith Huston	OARDC Wooster, OH 44691 (216) 263-0953
Program Coordinator	James H. Gibson	Natural Resource Ecology Laboratory Colorado State University Fort Collins, CO 80523 (303) 491-5571
Director--Central Analytical Laboratory	Gary Stensland	Illinois State Water Survey 605 East Springfield Champaign, IL 61820 (217) 333-2213
Representative--USDA	J.M. Barnes	Cooperative State Research Room 213 West Auditors Washington, DC 20250 (202) 447-5741
Representative--USGS	Jack Pickering	National Center, MS 412 2321 N. Richmond Street Arlington, VA 22207 (703) 860-6834
NCSU Acid Precipitation Program Coordinator	Ann Bartuska	North Carolina State University 1509 Varsity Drive Raleigh, NC 27606 (919) 737-3520

NATIONAL ATMOSPHERIC DEPOSITION PROGRAM/NATIONAL TRENDS NETWORK SITES

State	Code #	Network	Site Name	County	Agency	Individual	Lat.	Long.
AL (a)	011000	NTN	Blackbelt Substation	Dallas	USGS	I.A. Giles	32°27'	87°15'
AK (a)	020390	NADP	Denali (McKinley) Nat'l Pk.		NPS	J. Dalle Molle	63°43'	148°58'
AZ (d)	030360	NADP/NTN	Oliver Knoll	Graham	BLM	D. Molitor	33°04'	109°52'
AZ (c)	030370	NADP/NTN	Grand Canyon Nat'l Park		NPS	L. May	36°04'	112°09'
AZ (b)	030620	NADP/NTN	Organ Pipe Cactus, N.M.	Pima	NPS	W.F. Wallace	31°57'	112°48'
AR (b)	040260	NADP/NTN	Warren 2WSW	Bradley	NCASI-Potlatch	W.W. Pope	33°36'	92°06'
AR (d)	040380	NTN	Caddo Valley	Clark	USGS	C.T. Bryant	34°11'	93°06'
AR (c)	041620	NADP	Buffalo River	Marion	NPS	S. Chaney	36°05'	92°35'
AR (a)	042700	NADP/NTN	Fayetteville	Washington	SAES-UNIV. of Arkansas	G.H. Wagner	36°06'	94°10'
CA (f)	054200	NADP/NTN	Tanbark Flat	Los Angeles	USFS	P.R. Miller	34°12'	117°46'
CA (b)	054540	NADP/NTN	Hopland	Mendocino	BLM	N.Ritchey	39°00'	123°05'
CA (h)	056820	NADP	Palomar Mountain	San Diego	San Diego Gas & Electric	L.J. Brunton	33°19'	116°51'
CA (c)	057550	NADP/NTN	Sequoia National Park	Tulare	NPS	L. Bancroft	36°34'	118°47'
CA (i)	058501	NTN	Chuchupate Ranger Station	Ventura	USGS	E.W.Henrich	34°48'	119°01'
CA (a)	058840	NADP/NTN	Davis Site	Yolo	SAES-Univ. of California	R.H. Burgy	38°32'	121°46'
CA (g)	058850	NADP/NTN	Yosemite National Park	Tuolumne	NPS	R. Riegelhuth	37°47'	119°51'
CO (e)	060060	NADP/NTN	Alamosa	Alamosa	NOAA	J. Miller	37°27'	105°52'
CO (i)	060180	NTN	Las Animas Fish Hatchery	Bent	USGS	D. Cain	38°07'	103°14'
CO (g)	060220	NADP	Niwot Saddle	Boulder	Univ. Of Col.-INSTAAR	P.J. Webber	40°03'	105°35'
CO (f)	061530	NADP/NTN	Mesa Verde National Park	Montezuma	NPS	J.E. Smith	37°11'	108°29'
CO (b)	061560	NADP/NTN	Sand Spring	Moffat	BLM	O. Olsen	40°30'	107°42'
CO (d)	061910	NADP	Rocky Mountain Nat'l Park	Larimer	NPS	D.R. Stevens	40°22'	105°34'
CO (h)	061911	NADP	Loch Vale	Larimer	NPS	J. Baron	40°17'	105°39'
CO (j)	061920	NADP	Buffalo Pass	Routt	EPA/USFS	R. Bernardo	40°32'	107°41'
CO (a)	062120	NADP/NTN	Manitou	Teller	USFS	D.G. Fox	39°06'	105°06'
CO (c)	062220	NADP/NTN	Pawnee	Weld	SAES-Colorado State Univ.	J.H. Gibson	40°48'	104°45'
FL (a)	100360	NADP/NTN	Bradford Forest	Bradford	SAES-Univ of Florida	H. Riekerk	29°58'	82°12'
FL (d)	100380	NADP	CIFANT	Brevard	NASA/Bionetics Corp.	R. Hinkle	28°33'	80°39'
FL (c)	101190	NADP/NTN	Everglades National Park	Dade	NPS	P. Rosendahl	25°23'	80°42'

NATIONAL ATMOSPHERIC DEPOSITION PROGRAM/NATIONAL TRENDS NETWORK SITES

State	Code #	Network	Site Name	County	Agency	Individual	Lat.	Long.
FL (f)	101400	NTN	Quincy NTN Rainfall	Gadsden	USGS	G.A. Irwin	30°33'	84°36'
FL (e)	104100	NTN	Verna Well Field	Sarasota	USGS	D.B. Adams	27°23'	82°17'
GA (b)	112040	NADP	Bellville	Evans	Union Camp Corporation	F.S. Broerman	32°08'	81°58'
GA (a)	114140	NADP/NTN	Georgia Station	Pike	SAES - Univ. of Georgia	J.T. Walker	33°11'	84°24'
GA (c)	115000	NTN	Tifton, ARS	Tift	USGS/USDA-ARS	G.R. Buell	31°28'	83°32'
HI (a)	120080	NADP	Mauna Loa	Hawaii	NOAA	J. Miller	19°32'	155°35'
ID (a)	130340	NADP/NTN	Craters of the Moon	Butte	BLM/NPS	K. Gebhardt	43°28'	113°34'
ID (b)	130480	NADP/NTN	Headquarters	Clearwater	NCASI-Potlatch	D.J. McGreer	46°38'	115°49'
ID (c)	131180	NADP	Reynolds Creek	Owyhee	BLM/USDA-ARS	G.R. Stephen-son	43°12'	116°45'
ID (d)	131500	NADP/NTN	Smiths Ferry	Valley	Boise Cascade Corporation	V.J. Kollock	44°18'	116°04'
IL (a)	141160	NADP/NTN	Bondville	Champaign	SAES-Univ. of Illinois	G.J. Stensland	40°03'	88°22'
IL (d)	141800	NADP/NTN	NIARC	Dekalb	SAES-Univ. of Illinois	R. Bell	41°50'	88°51'
IL (e)	141980	NADP/NTN	Argonne	DuPage	Argonne Nat'l Laboratory	D.L. Sisterson	41°42'	87°59'
IL (b)	143580	NADP	SIU	Jackson	SAES-Southern Ill. Univ.	G.M. Aubertin	37°43'	89°16'
IL (f)	144740	NADP/NTN	Salem	Marion	NOAA/Barton ATC, Inc.	J. Miller	38°39'	88°58'
IL (c)	146340	NADP/NTN	Dixon Springs Ag. Ctr.	Pope	SAES-Univ. of Illinois	C.J. Kaiser	37° 26'	88°40'
IN (c)	152020	NTN	Huntington Reservoir	Huntington	USGS	W.E. Harkness	40°50'	85°28'
IN (d)	152260	NADP/NTN	Southwest Indiana	Knox	USGS/Purdue University	W.W. McFee	38°44'	87°29'
IN (a)	153421	NADP	IN Dunes Nat'l Lakeshore	Porter	NADP	J.R. Whitehouse	41°38'	87°05'
IN (b)	154100	NADP	Purdue Univ. Ag. Farm	Tippecanoe	SAES-Purdue University	K.T. Paw U	40°28'	86°59'
IA (b)	160880	NADP/NTN	Big Springs Fish Hatchery	Clayton	USGS/Iowa Conserv. Comm.	R. Buchmiller	42°54'	91°28'
IA (a)	162320	NADP/NTN	McNay Research Station	Lucas	USGS/IOWA State University	R. Buchmiller	40°57'	93°23'
KS (b)	170740	NTN	Farlington Fish Hatchery	Crawford	USGS	L.R. Shelton	37°38'	94°48'

NATIONAL ATMOSPHERIC DEPOSTITION PROGRAM/NATIONAL TRENDS NETWORK SITES

State	Code #	Network	Site Name	County	Agency	Individual	Lat.	Long.
KS (a)	173120	NADP/NTN	Konza Prairie	Riley	Kansas State University	L.C. Hulbert	39°06'	96°36'
KS (C)	173280	NADP/NTN	Lake Scott State Park	Scott	USGS/KANSAS St. Park Authority	C. Sauer	38°40'	100°48'
KY (b)	180360	NADP/NTN	Perryville	Boyle	U. Ky-Inst. Mining & Min Res.	D.W. Koppe- naal	37°41'	84°57'
KY (a)	182260	NADP/NTN	Lilley Cornett Woods	Letchner	NOAA/Eastern KY Univ.	D.D. Harding	37°05'	83°00'
KY (c)	183560	NTN	Clark State Fish Hatchery	Rowan	USGS/Moorehead State Univ.	D. Leist	38°07'	83°33'
KY (d)	183860	NADP/NTN	Land Between the Lakes	Trigg	TVA	W.J. Parkhurst	36°47'	88°04'
LA (B)	190620	NADP	Hill Farm Research Station	Clairborne	SAES-Louisiana State Univ.	S. Feagley	32°45'	93°03'
LA (a)	191260	NADP/NTN	Iberia Research Station	Iberia	USGS/SAES-LA State Univ.	S. Feagley	29°56'	91°43'
LA (C)	193060	NADP	Southeast Research Station	Washington	SAES-Louisiana State Univ.	S. Fealgey	30°48'	90°11'
ME (d)	200011	NADP/NTN	Acadia National Park	Hancock	ME Dep. Envir. Protection/ NPS	L. Carver	44°22'	68°16'
ME (b)	200045	NADP	Caribou	Aroostook	NOAA	J. Miller	46°52'	68°01'
ME (e)	200046	NADP/NTN	Presque Isle	Aroostook	NOAA/MAINE Ag. Exp. Station	J. Miller	46°39'	68°00'
ME (c)	200277	NADP/NTN	Bridgton	Cumberland	ME Dep. Env. Protection/ NPS	L. Carver	44°06'	70°40'
ME (a)	200935	NADP/NTN	Greenville Station	Piscataquis	SAES-UNIV. Of Maine	I.J. Fernandez	45°29'	69°40'
MD (b)	210360	NADP	White Rock	Carroll	Baltimore Gas & Elec. Co.	J.R. Lodge	76°58'	39°24'
MD (a)	211320	NADP/NTN	Wye	Queen Anne	SAES-Univ. of Maryland	R.B. Brins- field	38°55'	76°09'
MA (c)	220155	NADP	NACL	Barnstable	NPS	M.K. Foley	41°58'	70°01'
MA (b)	220815	NADP/NTN	Cadwell	Hampshire	SAES-Univ. of Mass.	D.L. Mader	42°21'	72°23'
MA (a)	221325	NADP/NTN	East	Middlesex	SAES-Univ. of Mass.	W.A. Feder	42°23'	71°13'

NATIONAL ATMOSPHERIC DEPOSITION PROGRAM/NATIONAL TRENDS NETWORK SITES

State	Code #	Network	Site Name	County	Agency	Individual	Lat.	Long.
MI (c)	230920	NADP/NTN	Douglas Lake	Cheboygan	SAES-Michigan State Univ.	R. Vandekopple	45°34'	84°41'
MI (g)	230980	NADP/NTN	Raco	Chippewa	EPA/USFS	R. Jewell	84°44'	46°22'
MI (f)	232241	NADP	Chassell	Houghton	NPA/Michigan St. Univ.	R. Stottle-myer	47°06'	88°33'
MI (d)	232570	NADP	Isle Royale National Park	Keweenow	NPS	R. Stottle-myer	47°54'	89°09'
MI (b)	232660	NADP/NTN	Kellogg Biological Station	Kalamazoo	SAES-Michigan St. Univ.	J.B. Hart, Jr.	42°25'	85°23'
MI (a)	235340	NADP/NTN	Wellston	Wexford	USFS	E.S. Verry	44°13'	85°49'
MN (a)	241660	NADP/NTN	Marcell Exp. Forest	Itasca	USFS	E.S. Verry	47°32'	93°28'
MN (c)	241840	NADP/NTN	Fernberg	Lake	EPA/USFS	G. Glass	47°57'	91°29'
MN (d)	242360	NTN	Camp Ripley	Morrison	USGS	M.R. Have	46°15'	94°29'
MN (b)	242720	NADP/NTN	Lamberton	Redwood	SAES-Un. of MN	W.W. Nelson	44°14'	95°18'
MS (b)	251080	NTN	Clinton	Hinds	USGS	G. Bednar	32°18'	90°18'
MS (a)	251460	NADP	Meridian	Lauderdale	NOAA	J. Miller	32°20'	88°45'
MS (c)	253000	NTN	Coffeeville	Yalobusha	USGS	G. Bednar	34°00'	89°48'
MO (a)	260380	NADP/NTN	Ashland Wildlife Area	Boone	SAES-Un. of MO	G.S. Henderson	38°45'	92°12'
MO (b)	260560	NADP/NTN	University Forest	Butler	SAES-Un. of MO	G.S. Henderson	36°55'	90°19'
MT (e)	270060	NADP/NTN	Custer Battlefield Nat'l Monument	Big Horn	USGS/NPS	J. Court	45°34'	107°26'
MT (c)	270560	NADP	St. Mary Ranger Station	Glacier	NPS	D. Lange	48°44'	113°26'
MT (a)	270570	NADP/NTN	Glacier Fire Weather	Flathead	NPS	D. Lange	48°30'	114°00'
MT (d)	270760	NTN	Clancy	Jefferson	USGS	R. Knapton	46°29'	112°03'
MT (b)	271340	NADP	Give Out Morgan	Roosevelt	EPA/Fort Peck Tribes	L. Allen	48°29'	105°12'
NE (a)	281520	NADP/NTN	Mead	Saunders	SAES-Un. of NE	S.B. Verma	41°09'	96°30'
NV (a)	290140	NADP/NTN	Saval Ranch	Elko	BLM/University of Nevada	K. Gebhardt	41°17'	115°49'
NH (a)	300240	NADP/NTN	Hubbard Brook	Grafton	USFS	J. Hornbeck	43°57'	71°42'
NJ (a)	312981	NADP/NTN	Washington Crossing	Mercer	NOAA/NJ Dep Env. Prot.	J. Miller	40°19'	74°51'
NM (b)	320720	NADP	Bandelier	Los Alamos	Los Alamos Nat'l Lab.	B. Ferenbaugh	35°47'	106°16'
NM (c)	320840	NADP	Mayhill	Otero	BLM/USFS	C. Anderson	32°54'	105°28'

NATIONAL ATMOSPHERIC DEPOSITION PROGRAM/NATIONAL TRENDS NETWORK SITES

State	Code #	Network	Site Name	County	Agency	Individual	Lat.	Long.
NM (a)	320980	NADP/NTN	CUBA	Sandoval	BLM	C. Anderson	36°02'	106°58'
NM (d)	321280	NADP	Capulin Mountain	Union	N.M. Environ. Improv. Div./ NPS	C. Eberhart	36°47'	103°58'
NY (c)	330860	NADP/NTN	Aurora Research Farm	Cayuga	SAES-Cornell Un.	W.W. Knapp	42°44'	76°40'
NY (g)	331000	NADP/NTN	Chautauqua	Chautauqua	Niagara Mohawk Power	R.M. Wood	42°18'	79°24'
NY (d)	331220	NADP/NTN	Knobit	Columbia	NY State Elec. & Gas	D. Matias	42°23'	73°30'
NY (a)	332020	NADP	Huntington Wildlife	Essex	St. U. of NY Syracuse	D.J. Raynal	43°58'	74°13'
NY (b)	335140	NADP	Stilwell Lake	Orange	U.S. Military Academy	R.C. Graham	41°21'	74°02'
NY (h)	335141	NADP/NTN	West Point	Orange	U.S. Military Academy	R.C. Graham	41°21'	74°03'
NY (f)	335240	NADP/NTN	Bennett Bridge	Oswego	Niagara Mohawk Power	R.M. Wood	43°32'	75°57'
NY (e)	336500	NADP/NTN	Jasper	Steuben	NY State Elec. & Gas	D. Matias	42°06'	77°32'
NY (i)	336840	NTN	Biscuit Brook Exp. Station	Ulster	USGS	P.S. Murdoch	42°00'	74°30'
NC (a)	340320	NADP/NTN	Lewiston	Bertie	SAES-NC State	R.I. Bruck	36°08'	77°10'
NC (b)	342500	NADP/NTN	Coweeta	Macon	USFS	J.E. Douglass	35°04'	83°26'
NC (c)	343460	NADP/NTN	Piedmont Research Station	Rowan	SAES-NC State	R.I. Bruck	35°42'	80°37'
NC (d)	343560	NADP/NTN	Clinton Crops Res. Station	Sampson	SAES-NC State	R.I. Bruck	35°01'	78°17'
NC (i)	343600	NTN	Jordan Creek	Scotland	USGS	J.K. Crawford	34°58'	79°31'
NC (e)	344160	NADP/NTN	Finley (a)	Wake	SAES-NC State	A.S. Heagle	35°44'	78°41'
ND (a)	350700	NADP/NTN	Theodore Roosevelt Nat'l Pk.	McKenzie	NPS	R. Powell	47°36'	103°16'
ND (c)	350880	NADP/NTN	Icelandic State Park	Pembina	USGS	R.L. Houghton	48°47'	97°45'
ND (b)	351180	NADP/NTN	Woodworth NTN	Slutsman	USGS	R.L. Houghton	47°07'	99°14'
OH (d)	360900	NADP/NTN	Oxford	Butler	USGS/Miami Un.	J.C. Klink	39°32'	84°43'
OH (a)	361760	NADP/NTN	Delaware	Delaware	USFS	L.S. Dochinger	40°21'	83°04'
OH (b)	364900	NADP/NTN	Caldwell	Noble	SAES-OH Ag. Res. & Dev. Ctr.	T.C. Weiden-saul	39°48'	82°32
OH (c)	367160	NADP/NTN	Wooster	Wayne	SAES-OH Ag. Res. & Dev. Ctr.	T.C. Weiden-saul	40°47'	81°56'
OK (c)	370060	NTN	Salt Plains Nat'l Wildl. Ref.	Alfalfa	USGS/U.S. Fish & Wildlife	J.H. Irwin	36°48'	98°12'
OK (b)	371740	NADP/NTN	Great Plains Apiaries	McClain	NOAA	E. Kessler	97°31'	34°59'

NATIONAL ATMOSPHERIC DEPOSITION PROGRAM/NATIONAL TRENDS NETWORK SITES

State	Code #	Network	Site Name	County	Agency	Individual	Lat.	Long.
OK (a)	372580	NADP	Clayton Lake	Pushmataha	SAES-OK St. Un.	P.J. Wigington, Jr.	34°32'	95°21'
OR (a)	380200	NADP	Alsea Guard Ranger Station	Benton	EPA	R.G. Wilhour	44°23'	123°37'
OR (j)	380202	NADP/NTN	Hyslop Farm	Benton	EPA	R.G. Wilhour	44°38'	123°11'
OR (g)	380260	NADP	Bull Run	Clackamas	USGS/City of Portland	S.W. McKenzie	45°27'	122°09'
OR (h)	380980	NTN	Silver Lake Ranger Station	Lake	USGS	J.F. Rinella	43°07'	121°04'
OR (f)	381020	NADP/NTN	H.J. Andrews Exp. Forest	Lane	USFS/Oregon St. Univ.	A. McKee	44°13'	122°15'
OR (c)	381120	NADP/NTN	Vines hill	Malheur	BLM	G. Ferry	43°54'	117°26'
OR (i)	381800	NTN	Starkey Exp. Forest	Union	USGS	J.F. Rinella	45°13'	118°30'
PA (c)	391520	NADP/NTN	Penn State	Centre	NOAA/Penn. St. Univ.	R.G. dePena	40°47'	77°57'
PA (a)	392940	NADP/NTN	Kane Experimental Forest	Elk	USFS	D.A. Marquis	41°36'	78°46'
PA (b)	394200	NADP/NTN	Leading Ridge	Huntingdon	SAES-Penn. St. Univ.	J.A. Lynch	40°40'	77°56'
PA (d)	397220	NADP/NTN	Milford	Pike	USFS/Gifford-Pinchott Inst.	E.S. Corbett	41°20'	74°49'
SC (b)	420680	NTN	Santee Nat'l Wildlife Refuge	Clarendon	USGS	G. Speiran	33°32'	80°26'
SC (a)	421880	NADP	Clemson	Pickens	SAES-Clemson Univ.	S.C. Hodges	34°40'	82°50'
SD (c)	430061	NTN	Huron Well Field	Beadle	USGS	W.L. Bradford	44°22'	98°17'
SD (b)	430880	NADP/NTN	Cottonwood	Jackson	NOAA/SD St. Univ.	J. Miller	43°57'	101°51'
TN (a)	440040	NADP/NTN	Walker Branch Watershed	Anderson	Oak Ridge Nat'l Lab.	S.E. Lindberg	35°58'	84°17'
TN (c)	441140	NADP/NTN	Giles County	Giles	TVA	W.J. Parkhurst	35°17'	86°54'
TN (b)	441190	NADP/NTN	Elkmont	Sevier	NPS	S. Coleman	35°40'	83°35'
TX (g)	450350	NADP/NTN	Beeville	Bee	NOAA/SAES-TX A & M	W.R. Ocumpaugh	27°27'	97°42'
TX (a)	450425	NADP/NTN	K-Bar	Brewster	NPS	M. Flemming	29°18'	103°11'
TX (k)	451040	NTN	Attwater Prairie Chicken NWR	Colorado	USGS	J. Rawson	29°15'	96°39'
TX (j)	451650	NADP/NTN	Sonora	Edwards	USGS/TX A&M AG. Exp. Sta.	C.S. Menzies	31°14'	100°19'
TX (d)	452180	NADP	Longview	Gregg	TX Air Control Board	S. Spaw	32°23'	94°43'

NATIONAL ATMOSPHERIC DEPOSITION PROGRAM/NATIONAL TRENDS NETWORK SITES

State	Code #	Network	Site Name	County	Agency	Individual	Lat.	Long.
TX (h)	452215	NADP/NTN	Guadalupe Mountains Nat'l Pk.	Culberson	USGS/NPS	K. Yarborough	31°54'	104°48'
TX (e)	453800	NADP/NTN	Forest Seed Center	Nacogdoches	Int. Paper Co.	W.G. Mumford	31°34'	94°52'
TX (i)	455180	NADP/NTN	Throckmorton	Throckmorton	USGS/Texas A&M	E.C. Gilmore	33°16'	99°12'
TX (b)	455350	NADP	Victoria	Victoria	NOAA	J. Miller	28°51'	96°55'
TX (f)	455640	NTN	LBJ National Grasslands	Wise	USGS	L. Harmsen	33°23'	97°38'
UT (b)	460120	NADP/NTN	Logan	Cache	NOAA/SAES-UT St. Univ.	G.L. Wool-dridge	41°39'	111°54'
VT (a)	470100	NADP/NTN	Bennington	Bennington	St. of VT/City of Bennington	R.L. Poirot	42°52'	73°10'
VA (a)	481300	NADP	Horton's Station	Giles	SAES-VPI & St. Univ.	B.I. Chevone	37°20'	80°33'
VA (b)	482890	NADP/NTN	Big Meadows		NPS	B.I. Chevone	38°31'	78°26'
VA (c)	483300	NADP/NTN	Loves Mill	Washington	TVA	W.J. Parkhurst	36°44'	81°41'
WA (a)	491410	NADP/NTN	Olympic Nat'l Park	Jefferson	NPS	B. Moorhead	47°52'	123°56'
WA (b)	491540	NADP	Sullivan Lake	Pend Oreille	Weyerhauser/NW Pulp-Paper/ USFS/WWU	D.F. Brakke	48°50'	117°17'
WA (d)	491940	NADP/NTN	Marblemount	Skagit	Weyerhauser/NW NPS/WWU	D.F. Brakke	48°31'	121°28'
WA (c)	492180	NADP	La Grande	Thurston	Weyerhauser/NW U WA/WWU	D.F. Brakke	46°45'	121°55'
WV (b)	500460	NTN	Babcock State Park	Fayette	USGS	T.A. Ehlke	37°59'	80°57'
WV (a)	501860	NADP/NTN	Parsons	Tucker	USFS	J.N. Kochen-derfer	39°05'	79°40'
WI (e)	512120	NADP	Legend Lake	Menominee	Wisc. Dep. of Nat. Res.	B.C. Rodger	44°53'	88°43'
WI (c)	512800	NADP/NTN	Lake Dubay	Portage	Nekoosa Paper/ Cons. Paper	D.C. Herman	44°40'	89°39'
WI (a)	513640	NADP/NTN	Trout Lake	Vilas	WI Dept. of Nat. Res.	R. Becker	46°03'	89°39'
WI (d)	513680	NADP/NTN	Lake Geneva	Walworth	Wisc. Dep. of Nat. Res	B.C. Rodger	42°35'	88°30'
WI (b)	513700	NADP/NTN	Spooner	Washburn	WI Dep. Nat. Res./ Univ. Wisc.	J.D. Chazin	45°49'	91°52'
WY (d)	520260	NADP	Sinks Canyon	Fremont	BLM	H. Oden	42°45'	108°48'
WY (c)	520680	NADP/NTN	Pinedale	Sublette	BLM	A.L. Riebau	42°56'	109°47'

NATIONAL ATMOSPHERIC DEPOSITION PROGRAM/NATIONAL TRENDS NETWORK SITES

State	Code #	Network	Site Name	County	Agency	Individual	Lat.	Long.
WY (b)	520820	NADP/NTN	Newcastle	Weston	BLM	A.L. Riebau	43°52'	104°12'
WY (a)	520860	NADP/NTN	Yellowstone National Park	Park	NPS	W. Hamilton	44°55'	110°25'
AS (a)	530190	NADP	Samoa		NOAA	J. Miller	14°15'	170°34'

SUBJECT INDEX . . . .

References are to item numbers, not page numbers